T0350183

WEAK CONVERGENCE AND ITS APPLICATIONS

WEAK CONVERGENCE AND ITS APPLICATIONS

Zhengyan Lin
Hanchao Wang
Zhejiang University, China

NEW JERSEY • LONDON • SINGAPORE • BEIJING • SHANGHAI • HONG KONG • TAIPEI • CHENNAI

Published by

World Scientific Publishing Co. Pte. Ltd.
5 Toh Tuck Link, Singapore 596224
USA office: 27 Warren Street, Suite 401-402, Hackensack, NJ 07601
UK office: 57 Shelton Street, Covent Garden, London WC2H 9HE

Library of Congress Cataloging-in-Publication Data
Lin, Zhengyan, 1941– author.
Weak convergence and its applications / Zhengyan Lin, Hanchao Wang, Zhejiang University, China.
pages cm
Includes bibliographical references and index.
ISBN 978-9814447690 (hardcover : alk. paper) -- ISBN 9814447692 (hardcover : alk. paper)
1. Stochastic processes. 2. Convergence. I. Wang, Hanchao, 1984– author. II. Title.
QA274.L538 2014
519.2'3--dc23
2014025623

British Library Cataloguing-in-Publication Data
A catalogue record for this book is available from the British Library.

Printed in Singapore

Preface

The book is devoted to the theory of weak convergence of probability measures on metric spaces.

There are many books concerning the weak convergence, for example, Ethier and Kurtz (1986), van der Vaart and Wellner (1996), Billingsley (1999), Jacod and Shiryaev (2003) and so on. The emphasis of these books are different. Statistics and econometrics have made great progress in last two decades. It became necessary to study the distributions or the asymptotic distributions of some complex statistics, so the weak convergence of stochastic processes based on these complex statistics would be more important. Some models may need new technique in weak convergence to study. For example,

• In the study of the asymptotic behavior of non-stationary time series, the limiting process (resp. limiting distribution) usually associates with the processes with conditional independent increment (resp. mixture normal distributions).

• The asymptotic errors caused by discretizations of stochastic differential equations (e.g. analysis of hedging error) converge to stochastic integral in distribution.

• In the finance risk theory, the returns of assets usually obey heavy-tailed distributions, the statistics based on such data do not have the asymptotic normality. The processes based on these statistics usually converge to the non-Gaussian stable process weakly.

• Some statistics in the goodness-of-fit testing are empirical processes or functional index empirical processes. It is quite difficult to state the weak convergence of such processes.

Our main aim is to give a systematic exposition of the theory of weak convergence, covering wide range of weak convergence problems including new developments. In this book, we summarize the development of weak convergence in each of the following aspects: Donsker type invariance principles, convergence of point processes, weak convergence to semimartingale and convergence of empirical processes. Compared to the well-known monographs mentioned earlier, we do not cover all details, but our book contains the core content of many branches and new development as much as possible. This is perhaps one of the highlights of our book.

In chapter 1, some definitions, properties of weak convergence are presented, including the useful metric space, portmanteau theorem of weak convergence and some important examples. It is noteworthy that the method of proving weak convergence are given, which is frequently used in the following chapter.

The weak convergence to independent increment processes is presented in chapter 2. We divide this chapter to two main parts: Donsker type invariance principles and point process convergence. Donsker type invariance principles depict weak convergence to Gaussian independent increment processes, and point process convergence is about the weak convergence to compound Poisson type processes.

In chapter 3, We give the weak convergence to semimartingale. Firstly, the general case of convergence to semimartingale is presented. Furthermore, weak convergence to stochastic integral as examples is given. At last, some applications are collected.

Weak convergence of empirical processes are presented in chapter 4. The classical results of empirical process are presented in section 4.1, and the convergence of function index empirical processes are discussed in section 4.3 and 4.4. Section 4.2 and 4.5 give some applications of weak convergence of empirical processes.

More and more people are interested in applications of weak convergence theory. Another aim of writing this book is to review some recent applications of modern weak convergence theory in time series, statistics and econometrics. Some of the results belong to the authors.

Finally, we would like to sincerely thank Prof. V. Ulyanov of Moscow State University, who provided us valuable comments.

Contents

Chapter 1

The Definition and Basic Properties of Weak Convergence

1.1 Metric space

The aim of writing this book is to summarize both classical theory and recent development of weak convergence theory. The weak convergence of stochastic processes is an important subject of modern probability theory. Stochastic processes can be treated as the random elements taking values in a specific space, such as function space. To study the weak convergence of stochastic processes, the space, in which the stochastic processes take values, should be a *Polish space* (a complete separable topological space) with a suitable topology. Usually, people have been interested in the topology of the space. In this section, we firstly review the properties of some useful topological spaces, and the definitions of weak convergence in these spaces.

The following notations would be used in the whole book.

Let $(\Omega, \mathcal{F}, \mathbb{P})$ be the probability space, X the random element on $(\Omega, \mathcal{F}, \mathbb{P})$ taking values in the space $\mathbb{S}$. $\mathcal{B}(\mathbb{S})$ stands for the $\sigma-$field generated by open subsets of $\mathbb{S}$ under the metric ρ. $(\mathbb{S}, \rho)$ stands for the metric space. In this book, we mainly focus on three kinds of spaces for the study of weak convergence: $\mathbb{C}[0,1]$, $\mathbb{D}[0,1]$ and $\mathbb{M}_+(\mathbb{S})$. $\mathbb{R}$ stands for the 1-dimensional Euclid space.

1.1.1 $\mathbb{C}[0,1]$ *space*

$\mathbb{C}[0,1]$ is the space of all continuous functions: $[0,1] \to \mathbb{R}$. We wish to introduce a topology to $\mathbb{C}[0,1]$, such that $\mathbb{C}[0,1]$ is a Polish space with this topology. Usually, for $\alpha, \beta \in \mathbb{C}[0,1]$, people use the *local uniform topology* associated with the metric

$$\rho(\alpha, \beta) = \sup_{s \in [0,1]} |\alpha(s) - \beta(s)|. \tag{1.1}$$

By the *Arzelà-Ascoli theorem*, a set $A \subset \mathbb{C}[0,1]$ is relatively compact if and only if

$$\sup_{t \in [0,1]} \sup_{x \in A} |x(t)| < \infty$$

and

$$\lim_{\delta \downarrow 0} \sup_{x \in A} \sup_{|t-s| \leq \delta} |x(t) - x(s)| = 0.$$

1.1.2 $\mathbb{D}[0,1]$ *space*

$\mathbb{D}[0,1]$ is the space of all functions: $[0,1] \to \mathbb{R}$, which are right-continuous and have left-hand limits.

Since the elements in $\mathbb{D}[0,1]$ may have discontinuous points, it is difficult to find a topology such that $\mathbb{D}[0,1]$ is a Polish space endowed this topology.

For briefness and convenience, people may hope that $\mathbb{D}[0,1]$ will be a Polish space with the local uniform topology. Unfortunately, $\mathbb{D}[0,1]$ fails to be separable with this metric. For example, the number of functions

$$\alpha_s(t) = 1_{[s,1)}(t), \quad 0 \le s \le 1$$

is uncountable. However, $\rho(\alpha_s, \alpha_{s'}) = 1$ for $s \neq s'$, so $\mathbb{D}[0,1]$ is not separable with the local uniform topology.

In fact, $\mathbb{D}[0,1]$ is uncountable under the uniform topology because of the discontinuity of $x(t)$, a function belonging to $\mathbb{D}[0,1]$. The candidate topology should be different. *Skorokhod (1956)* introduced four kinds of topologies to overcome the shortage of the local uniform topology.

To express the definitions of these four topologies briefly, we use the convergence concept to define the topologies.

Definition 1.1. Let x_n, $n \ge 1$, x be elements in $\mathbb{D}[0,1]$. If there exists a continuous one-to-one mapping sequence $\lambda_n(t)$ of $[0,1]$ onto $[0,1]$ such that

$$\lim_{n\to\infty} \sup_{t\in[0,1]} |x_n(\lambda_n(t)) - x(t)| = 0, \qquad \lim_{n\to\infty} \sup_{t\in[0,1]} |\lambda_n(t) - t| = 0, \tag{1.2}$$

we call x_n J_1-convergent to x, denoted by $x_n \xrightarrow{J_1} x$.

The definition of J_1 *topology* can be introduced by J_1*-convergence.* Usually, J_1 topology stands for *Skorokhod topology*, but there are many situations where J_1 should be replaced by the other kinds of topologies.

Definition 1.2. Let x_n, $n \ge 1$, x be elements in $\mathbb{D}[0,1]$, and if there exists a one-to-one mapping sequence $\lambda_n(t)$ of $[0,1]$ onto $[0,1]$ such that (1.2) is satisfied, we call x_n J_2-convergent to x, denoted by $x_n \xrightarrow{J_2} x$.

Define the *graph* $\Gamma_{x(t)}$ as the closed set in $\mathbb{R} \times [0,1]$, such that for any t, $x(t)$ belongs to $[x(t-0), x(t+0)]$.

If there exists a pair of functions $(y(s), t(s))$, and an s such that $x = y(s)$, $t = t(s)$, where $y(s)$ is continuous, $t(s)$ is continuous and monotonically increasing, those and only those pairs (x,t) belong to $\Gamma_{x(t)}$, we call $(y(s), t(s))$ a *parametric representation* of $\Gamma_{x(t)}$.

We use the above notations and definitions to define the M_1 and M_2 topologies. Let $R[(x_1,t_1);(x_2,t_2)] = |x_1 - x_2| + |t_1 - t_2|$.

Definition 1.3. If there exist parametric representations $(y(s), t(s))$ of $\Gamma_{x(t)}$ and $(y_n(s), t_n(s))$ of $\Gamma_{x_n(t)}$ such that

$$\lim_{n\to\infty} \sup_{s\in[0,1]} R[(y_n(s), t_n(s)); (y(s), t(s))] = 0, \tag{1.3}$$

we call x_n M_1-convergent to x, denoted by $x_n \xrightarrow{M_1} x$.

Definition 1.4. If there exist parametric representations $(y(s), t(s))$ of $\Gamma_{x(t)}$ and $(y_n(s), t_n(s))$ of $\Gamma_{x_n(t)}$ such that

$$\lim_{n\to\infty} \sup_{(y_1,t_1)\in\Gamma_{x(t)}} \inf_{(y_2,t_2)\in\Gamma_{x_n(t)}} R[(y_1, t_1); (y_2, t_2)] = 0, \tag{1.4}$$

we call x_n M_2-convergent to x, denoted by $x_n \xrightarrow{M_2} x$.

The definitions of M_1 and M_2 topologies which are given in terms of graphs are actually complicated and not short. We will present the necessary and sufficient conditions for M_1 and M_2 convergence.

Set $H(x_1, x_2, x_3)$ is the distance of x_2 from $[x_1, x_3]$.

Proposition 1.1. *(Skorohod (1956))*

There is equivalence between:

(i) $x_n \xrightarrow{M_1} x$*;*

(ii) a) x_n *converges to* x *on a dense set of* $[0, 1]$, *and this set contains 0 and 1.*

b)

$$\lim_{c\to 0} \limsup_{n\to\infty} \sup_{t_2-c<t_1<t_2<t_3<t_2+c} H(x_n(t_1), x_n(t_2), x_n(t_3)) = 0.$$

Proposition 1.2. *(Skorohod (1956))*

There is equivalence between:

(i) $x_n \xrightarrow{M_2} x$*;*

(ii)

$$\lim_{c\to 0} \limsup_{n\to\infty} \sup_{t\in[0,1]} H(x(t), x_n(t_c), x(t_c^*)) = 0,$$

where $t_c = \max\{0, t-c\}$, $t_c^* = \min\{1, t+c\}$.

Next, we present the conditions for compactness in the topologies J_1, J_2, M_1, M_2.

Proposition 1.3. *(Skorohod (1956))*

The set K *is compact in a topology* S, *where* S *stands for one of* J_1, J_2, M_1, M_2, *if*

$$\lim_{c\to 0} \lim_{x\in K} \sup_{t\in[0,1]} (\Delta_S(c, x(t)) + \sup_{0<t<c} |x(t) - x(0)| + \sup_{1-c<t<1} |x(t) - x(1)|) = 0,$$

where

$$\Delta_{J_1}(c, x(t)) = \sup_{t-c<t_1<t\le t_2<t+c,\ t_1,t_2\in[0,1]} \min\{|x(t_1) - x(t)|, |x(t_2) - x(t)|\},$$

$$\Delta_{J_2}(c, x(t)) = \sup_{t_c \le t_1 \le t_c + c/2,\ t, t_1, t_2 \in [0,1],\ t_c - c/2 \le t_2 \le t_c} \min\{|x(t_1) - x(t)|, |x(t_2) - x(t)|\},$$

$$\Delta_{M_1}(c, x(t)) = \sup_{t_2 - c < t_1 < t_2 < t_3 < t_2 + c,\ t_1, t_2, t_3 \in [0,1]} H(x(t_1), x(t_2), x(t_3)),$$

$$\Delta_{M_2}(c, x(t)) = \sup_{t_c \le t_1 \le t_c + c/2,\ t, t_1, t_2 \in [0,1],\ t_c^* - c/2 \le t_2 \le t_c^*} H(x(t), x(t_1), x(t_2)).$$

Usually, when the limiting process takes value in $\mathbb{D}[0,1]$, people study the asymptotic properties under the J_1 and M_1 topologies. However, the J_1 and M_1 topologies are not enough for convergence in some special cases.

Example 1.1. Let $k^0(t) = 1_{[1/2,1]}(t)$, $k^n(t) = 1_{[1/2-1/n,1]}(t)$, $x^0(t) = x^n(t) = 1_{[1/2,1]}(t)$, where $n \ge 1$. Then $x^n \xrightarrow{J_1} x^0$, $k^n \xrightarrow{J_1} k^0$, however $x^0(s) \equiv \int_0^1 k^n(s-)dx^n(s)$ doesn't converge to $\int_0^1 k^0(s-)dx^0(s) \equiv 0$ in J_1 topology. It means that the convergence of integrators and integrands in J_1 topology can not implies the convergence of integral in J_1 topology.

Example 1.2. Let $k^0(t) = 1_{[1/2,1]}(t)$, $k^n(t) = 1_{[1/2+1/n,1]}(t)$, $x^0(t) = x^n(t) = 1_{[1/2,1]}(t)$, where $n \ge 1$. Then $x^n \xrightarrow{J_1} x^0$, $k^n \xrightarrow{J_1} k^0$,

$$0 \equiv \int_0^1 k^n(s-)dx^n(s) \xrightarrow{J_1} \int_0^1 k^0(s-)dx^0(s) \equiv 0,$$

however, $(k^n, x^n) \xrightarrow{J_1} (k^0, x^0)$ doesn't hold. It means that the convergence of integral in J_1 topology can not imply joint convergence of integrators and integrands in J_1 topology.

Although Skorohod topology is very common in the study of weak convergence theory, it is not suitable when we study the convergence of some functionals in $\mathbb{D}[0,1]$. The examples 1.1 and 1.2 tell us that convergence of integrals has little contact with convergence of integrands and integrators under Skorohod topology. In *Jakubowski (1996, 1997)*, the author introduced a new topology, the so-called *semimartingale topology*. Although this topology can not be metrical, it is quite natural and share many useful properties with J_1 and M_1. Furthermore, it is advantage for convergence of integrals, we will present the details in Chapter 3. In this section, we firstly give the definition of the topology.

Definition 1.5. If x_n, x belong to $\mathbb{D}[0,1]$, and for every $\varepsilon > 0$, one can find bounded variation functions $v_{n,\varepsilon}$ and v_ε, such that

$$\sup_{t\in[0,1]} |x_n(t) - v_{n,\varepsilon}(t)| \le \varepsilon,$$

$$\sup_{t\in[0,1]} |x(t) - v_\varepsilon(t)| \le \varepsilon,$$

and for every continuous function $f : [0,1] \to \mathbb{R}$,

$$\int_{[0,1]} f(t)dv_{n,\varepsilon}(t) \to \int_{[0,1]} f(t)dv_\varepsilon(t).$$

we call x_n S-convergent to x, denoted by $x_n \xrightarrow{S} x$.

To characterize the compact set under the S topology in $\mathbb{D}[0,1]$, we need the following notation.

For a function $k \in \mathbb{D}[0,1]$, let $N_\eta(k)$ be the number of $\eta-$oscillations of k in $[0,1]$. More precisely, $N_\eta(k) \geq m$ if there are points $0 \leq t_1 \leq t_2 \leq \cdots \leq t_{2m-1} \leq t_{2m} \leq 1$, such that $|k(t_{2j}) - k(t_{2j-1})| > \eta$, $j = 1, 2, \cdots, m$.

Proposition 1.4. *$K \subset \mathbb{D}[0,1]$ is $S-$compact if and only if for every $\eta > 0$, there exists constant C_η such that*

(i)

$$\sup_{k \in K} \sup_{t \in [0,1]} |k(t)| \leq C_\eta,$$

(ii)

$$\sup_{k \in K} N_\eta(k) \leq C_\eta.$$

The proof of this proposition can be found in Skorohod (1956).

1.1.3 $\mathbb{M}_+(\mathbb{S})$ *space*

In some stochastic models, people usually define the stochastic point process to be a counting function for modeling random distribution of points in a locally compact topological space $\mathbb{S}$. A stochastic point process can be seen as a random element taking values in the measure space, then it is necessary to discuss the measure space's topology for studying the weak convergence of point processes.

Suppose that $\mathbb{S}$ is a locally compact topological space with a countable base. $\mathcal{B}(\mathbb{S})$ stands for the $\sigma-$field generated by open subsets of $\mathbb{S}$.

A measure μ on $(\mathbb{S}, \mathcal{B}(\mathbb{S}))$ is called Radon if $\mu(K) < \infty$ for every compact set $K \subset \mathbb{S}$.

Define

$$\mathbb{M}_+(\mathbb{S}) = \{\mu : \mu \text{ is Radon measure on } (\mathbb{S}, \mathcal{B}(\mathbb{S}))\},$$

$$\mathbb{C}_b^+(\mathbb{S}) = \{f : \mathbb{S} \mapsto \mathbb{R}_+; \text{ is continuous with compact support}\}.$$

Definition 1.6. If $\mu_n \in \mathbb{M}_+(\mathbb{S})$, $n \geq 0$, and for every $f \in \mathbb{C}_b^+(\mathbb{S})$,

$$\int_{\mathbb{S}} f d\mu_n \to \int_{\mathbb{S}} f d\mu_0$$

as $n \to \infty$, we call μ_n converges vaguely to μ_0, denoted by $\mu_n \xrightarrow{v} \mu_0$.

Vague topology can be introduced by *vague convergence.*

A very important element in $\mathbb{M}_+(\mathbb{S})$ is the point measure with the form

$$\mu = \sum_i \delta_{x_i}$$

where δ denotes the Dirac measure, $\{x_i\}$ is a sequence of points in $\mathbb{S}$.

Let $\mathbb{M}_p(\mathbb{S})$ be the set of all Radon point measures in $\mathbb{M}_+(\mathbb{S})$. $\mathbb{M}_p(\mathbb{S})$ turns out to be a closed subset of $\mathbb{M}_+(\mathbb{S})$.

To specify open sets and a topology in $\mathbb{M}_+(\mathbb{S})$, we need to define a basis set to be a subset of $\mathbb{M}_+(\mathbb{S})$ with the form

$$\{\mu \in \mathbb{M}_+(\mathbb{S}) : \int_{\mathbb{S}} f_i d\mu \in (a_i, b_i), i = 1, 2, \cdots, d\},$$

where d is any fixed positive integer and $f_i \in \mathbb{C}_b^+(\mathbb{S})$, and $0 \le a_i \le b_i$. Unions of basis sets form the class of open sets constituting the vague topology. Furthermore, let $\{f_i\}$ be the countable basis of $\mathbb{C}_b^+(\mathbb{S})$, for μ_1 and $\mu_2 \in \mathbb{M}_+(\mathbb{S})$, define metric

$$d(\mu_1, \mu_2) = \sum_{i=1}^{\infty} \frac{|\int_S f_i d\mu_1 - \int_S f_i d\mu_2| \wedge 1}{2^i}.$$

Thus, vague topology is metrizable as a complete, separable metric space. The following proposition characterize the relatively compact set in $\mathbb{M}_+(\mathbb{S})$

Proposition 1.5. *(Resnick (2007)) A set $M \subset \mathbb{M}_+(\mathbb{S})$ is vaguely relatively compact if and only if*

$$\sup_{\mu \in M} \int_{\mathbb{S}} f d\mu < \infty$$

for every $f \in \mathbb{C}_b^+(\mathbb{S})$.

1.2 The definition of weak convergence of stochastic processes and portmanteau theorem

The weak convergence of stochastic processes is an important subject. In fact, the weak convergence of stochastic processes is the weak convergence of probability measures, which is defined on the functional space. So firstly we give the definition of the weak convergence of probability measures. In this section, $(\mathbb{S}, d)$ stands for a complete separable space, $\mathcal{G}(\mathbb{S})$, $\mathcal{F}(\mathbb{S})$ and $\mathcal{K}(\mathbb{S})$ stand for the families of open sets, closed sets and compact sets respectively in $\mathbb{S}$. $\mathcal{P}(\mathbb{S})$ stands for the family of probability measures defined on the $\mathbb{S}$. $C(\mathbb{S})$ stands for the space of real-valued bounded continuous functions on the $(\mathbb{S}, d)$.

In the previous section, we have already defined the topology for the $\mathbb{M}_+(\mathbb{S})$. When people study the special case $\mathcal{P}(\mathbb{S})$ of $\mathbb{M}_+(\mathbb{S})$, the vague convergence is equivalence to the weak convergence, thus convergence in $\mathcal{P}(\mathbb{S})$ is very interesting. In this section, we will first introduce the Prohorov metric in $\mathcal{P}(\mathbb{S})$. Then we will present the connection between Prohorov metric, vague topology and weak convergence.

1.2.1 *Prohorov metric*

Prohorov (1956) introduced the *Prohorov metric* as follow:

$$\rho(\mathbb{P},\mathbb{Q}) = \inf\{\varepsilon > 0 : \ \mathbb{P}(F) \leq \mathbb{Q}(F^{\varepsilon}) + \varepsilon \text{ for every } F \in \mathcal{F}(\mathbb{S})\}, \tag{1.5}$$

where $\mathbb{P},\mathbb{Q} \in \mathcal{P}(\mathbb{S})$, $F^{\varepsilon} = \{x \in \mathbb{S} : \inf_{y\in F} d(x,y) < \varepsilon\}$.

Proposition 1.6. *ρ is a metric on $\mathcal{P}(\mathbb{S})$.*

Proof. Firstly, if $\rho(\mathbb{P},\mathbb{Q}) = 0$, then for every $F \in \mathcal{F}(\mathbb{S})$, $\mathbb{P}(F) = \mathbb{Q}(F)$, hence for every $A \in \mathcal{B}(\mathbb{S})$, $\mathbb{P}(A) = \mathbb{Q}(A)$. Therefore, $\rho(\mathbb{P},\mathbb{Q}) = 0$ if and only if $\mathbb{P} = \mathbb{Q}$.

Furthermore, for every $F \in \mathcal{F}(\mathbb{S})$, there exist $\alpha, \beta > 0$ such that

$$\mathbb{P}(F) \leq \mathbb{Q}(F^{\alpha}) + \beta.$$

For any $K \in \mathcal{F}(\mathbb{S})$, let $F = \mathbb{S} - K^{\alpha}$, then $F \in \mathcal{F}(\mathbb{S})$, and $K \subset \mathbb{S} - F^{\alpha}$, then

$$\mathbb{P}(K^{\alpha}) = 1 - \mathbb{P}(F) \geq 1 - \mathbb{Q}(F^{\alpha}) - \beta \geq \mathbb{Q}(K) - \beta,$$

which implies

$$\mathbb{Q}(K) \leq \mathbb{P}(K^{\alpha}) + \beta$$

for every $K \in \mathcal{F}(\mathbb{S})$. We have $\rho(\mathbb{P},\mathbb{Q}) = \rho(\mathbb{Q},\mathbb{P})$.

Finally, if $\mathbb{T} \in \mathcal{P}(\mathbb{S})$, $\rho(\mathbb{P},\mathbb{Q}) < \delta$, $\rho(\mathbb{Q},\mathbb{T}) < \varepsilon$, then for every $F \in \mathcal{F}(\mathbb{S})$,

$$\mathbb{P}(F) \leq \mathbb{Q}(F^{\delta}) + \delta \leq \mathbb{Q}(\overline{F^{\delta}}) + \delta \leq \mathbb{T}((\overline{F^{\delta}})^{\varepsilon}) + \delta + \varepsilon = \mathbb{T}(F^{\delta+\varepsilon}) + \delta + \varepsilon,$$

so $\rho(\mathbb{P},\mathbb{T}) \leq \delta + \varepsilon$. It implies $\rho(\mathbb{P},\mathbb{T}) \leq \rho(\mathbb{P},\mathbb{Q}) + \rho(\mathbb{Q},\mathbb{T})$. The proposition is obtained. □

The following theorem tells us the reason why $\mathbb{S}$ is to be a complete separable space.

Theorem 1.1. *$(\mathcal{P}(\mathbb{S}),\rho)$ is a complete separable space.*

Proof. Let $\{x_n\}$ be a countable dense subset of $\mathbb{S}$. Let δ_x denote the indicator measure of point x, it obviously belongs to $\mathcal{P}(\mathbb{S})$. Let $\{a_i\}_{i=1}^{\infty}$ denote the rational number set and N denote the finite positive integer. By the monotone class theorem and the convergence theorem, the probability measures of the form $\sum_{i=1}^{N} a_i \delta_{x_i}$ comprise a dense subset of $\mathcal{P}(\mathbb{S})$. We get the separability of $(\mathcal{P}(\mathbb{S}),\rho)$.

For completeness, we consider the sequence $\{\mathbb{P}_n\} \subset \mathcal{P}(\mathbb{S})$ with $\rho(\mathbb{P}_n,\mathbb{P}_{n-1}) < 2^{-n}$ for each $n \geq 2$. We need to find a $\mathbb{P} \in \mathcal{P}(\mathbb{S})$, such that $\rho(\mathbb{P}_n,\mathbb{P}) \to 0$ as $n \to \infty$.

The basic idea of proof is to construct a probability space $(\Omega,\mathcal{F},\nu)$ and $\mathbb{S}$−valued random variables X, X_n, $n = 1,2,\cdots$, with distribution $\mathbb{P}$ and $\mathbb{P}_n$. If we already have $d(X_n,X) \to 0$ in probability as $n \to \infty$ due to the completeness of $(\mathbb{S},d)$, we can obtain that for every $\varepsilon > 0$,

$$\lim_{n\to\infty} \mu_n((x,y) : d(x,y) \geq \varepsilon) = 0,$$

where μ_n is the joint distribution of X_n and X. Then, for every $F \in \mathcal{F}(\mathbb{S})$,

$$\begin{aligned}\mathbb{P}_n(F) &= \mu_n(F \times \mathbb{S})\\ &\leq \mu_n((F \times \mathbb{S}) \cap \{(x,y) : d(x,y) \leq \varepsilon\}) + \delta\\ &\leq \mu_n(\mathbb{S} \times F^\varepsilon) + \delta = \mathbb{P}(F^\varepsilon) + \delta,\end{aligned}$$

for given $\delta > 0$ and large n. Hence $\rho(\mathbb{P}_n, \mathbb{P}) \to 0$ as $n \to \infty$.

Now, we construct the probability space and random variables. Since $\rho(\mathbb{P}_n, \mathbb{P}_{n-1}) \leq 2^{-n}$, we first construct $(\Omega, \mathcal{F}, \nu)$, X_n and X_{n-1}, such that

$$\nu(d(X_n, X_{n-1}) \geq 2^{-n+1}) \leq 2^{-n+1},$$

X_{n-1}, X_n have distributions $\mathbb{P}_{n-1}$, $\mathbb{P}_n$ respectively.

Firstly, choose $E_1^{(n)}, \cdots, E_{N_n}^{(n)} \in \mathcal{B}(\mathbb{S})$ disjoint with diameters less than 2^{-n}, and $\mathbb{P}_{n-1}(E_0^{(n)}) \leq 2^{-n}$, where $E_0^{(n)} = \mathbb{S} - \cup_{i=1}^{N_n} E_i^{(n)}$.

Let $p_{n,i} = \mathbb{P}_{n-1}(E_i^{(n)})$, $A_i^{(n)} = (E_i^{(n)})^{2^{-n}}$, $i = 1, \cdots, N_n$. We have proved

$$\sum_{i\in I} p_{n,i} \leq \mathbb{P}_{n-1}(\overline{\bigcup_{i\in I} E_i^{(n)}}) \leq \mathbb{P}_n(\bigcup_{i\in I} A_i^{(n)}) + 2^{-n} \tag{1.6}$$

for all $I \subset \{1, 2, \cdots, N_n\}$. In the following, we will prove that there exist positive Borel measures $\eta_1^n, \eta_2^n, \cdots, \eta_{N_n}^n$ on $\mathbb{S}$ such that

$$\begin{cases}\eta_i^n(A_i^{(n)}) = \eta_i^n(\mathbb{S}) \leq p_{n,i} \quad \text{for } i = 1, 2, \cdots, N_n,\\ \sum_{i=1}^{N_n} \eta_i^n(\mathbb{S}) \geq \sum_{i=1}^{N_n} p_{n,i} - 2^{-n}\\ \sum_{i=1}^{N_n} \eta_i^n(A) \leq \mathbb{P}_n(A) \text{ for } \ A \in \mathcal{B}(\mathbb{S}).\end{cases} \tag{1.7}$$

We claim that (1.6) can be replaced by

$$\sum_{i\in I} p_{n,i} \leq \mathbb{P}_{n-1}(\overline{\bigcup_{i\in I} E_i^{(n)}}) \leq \mathbb{P}_n(\bigcup_{i\in I} A_i^{(n)}). \tag{1.8}$$

We will prove that there exist positive Borel measures $\eta_1^n, \eta_2^n, \cdots, \eta_{N_n}^n$ on $\mathbb{S}$ such that

$$\begin{cases}\eta_i^n(A_i^{(n)}) = \eta_i^n(\mathbb{S}) = p_{n,i} \quad \text{for } \ i = 1, 2, \cdots, N_n,\\ \sum_{i=1}^{N_n} \eta_i^n(A) \leq \mathbb{P}_n(A) \ \text{ for all } A \in \mathcal{B}(\mathbb{S}).\end{cases} \tag{1.9}$$

In fact, we proceed by induction on N_n. For $N_n = 1$, define $\eta_1^n(A) = p_{n,1}\mathbb{P}_n(A \cap A_1^{(n)})/\mathbb{P}_n(A_1^{(n)})$ for $A \in \mathcal{B}(\mathbb{S})$. Obviously, η_1^n satisfies the requirement.

Suppose that the argument holds for $m = 1, 2, \cdots, N_n - 1$. Define $\lambda(A) = \mathbb{P}_n(A \cap A_{N_n}^{(n)})/\mathbb{P}_n(A_{N_n}^{(n)})$, and let δ_0 be the largest δ such that

$$\sum_{i\in I} p_{n,i} \leq (\mathbb{P}_n - \delta\lambda)(\bigcup_{i\in I} A_i^{(n)}) \tag{1.10}$$

for all $I \subset \{1, 2, \cdots, N_n - 1\}$.

When $\delta_0 \geq p_{n,N_n}$, let $\eta_{N_n}^{(n)} = p_{n,N_n}\lambda$, and $\mathbb{P}'_n = \mathbb{P}_n - \eta_{N_n}^{(n)}$. Since the induction hypothesis, there exist positive Borel measures $\eta_1^{(n)}, \eta_2^{(n)}, \cdots, \eta_{N_n-1}^{(n)}$ on $\mathbb{S}$ such that

$\eta_i^{(n)}(A_i^{(n)}) = \eta_i^{(n)}(\mathbb{S}) = p_{n,i}$ for $i = 1, 2, \cdots, N_n - 1$, and $\sum_{i=1}^{N_n-1} \eta_i^{(n)}(A) \leq \mathbb{P}'_n(A)$ for all $A \in \mathcal{B}(\mathbb{S})$, and $\eta_{N_n}^{(n)}(A_{N_n}^{(n)}) = \eta_{N_n}^{(n)}(\mathbb{S}) = p_{n,N_n}$, the argument holds.

When $\delta_0 < p_{n,N_n}$, let $\mathbb{P}''_n = \mathbb{P}_n - \delta_0\lambda$. By the definition of δ_0, there exists $I_0 \subset \{1, 2, \cdots, N_n - 1\}$, such that

$$\sum_{i \in I} p_{n,i} \leq \mathbb{P}''_n(\bigcup_{i \in I} A_i^{(n)})$$

for all $I \subset I_0$. By the induction hypothesis, there exist positive Borel measures $\eta_i^{(n)}$ on $\mathbb{S}$ such that $\eta_i^{(n)}(A_i^{(n)}) = \eta_i^n(\mathbb{S}) = p_{n,i}$ for $i \in I_0$, and $\sum_{i \in I_0} \eta_i^n(A) \leq \mathbb{P}''_n(A)$ for all $A \in \mathcal{B}(\mathbb{S})$. Let $p'_{n,i} = p_{n,i}$ for $i = 1, 2, \cdots, N_n - 1$, $p'_{n,N_n} = p_{n,N_n} - \delta_0$, $B_0 = \bigcup_{i \in I_0} A_i^{(n)}$. Define $\widetilde{\mathbb{P}}_n(A) = \mathbb{P}''_n(A) - \mathbb{P}''_n(A \cap B_0)$, $I_1 = \{1, 2, \cdots, n\} - I_0$.

Then for every $I \subset I_1$

$$\begin{aligned}
\sum_{i \in I} p'_{n,i} + \mathbb{P}''_n(B_0) &= \sum_{i \in I \bigcup I_0} p'_{n,i} \leq \mathbb{P}''_n(\bigcup_{i \in I \bigcup I_0} A_i^{(n)}) \\
&= \mathbb{P}''_n(\bigcup_{i \in I} A_i^{(n)}) + \mathbb{P}''_n(B_0^n) - \mathbb{P}''_n(\bigcup_{i \in I} A_i^{(n)} \cap B_0^{(n)}) \\
&= \widetilde{\mathbb{P}}_n(\bigcup_{i \in I} A_i^{(n)}) + \mathbb{P}''_n(B_0^{(n)}).
\end{aligned}$$

Obviously,

$$\sum_{i \in I} p'_{n,i} \leq \widetilde{\mathbb{P}}_n(\bigcup_{i \in I} A_i^{(n)})$$

for all $I \subset I_1$. Since the induction hypothesis, there exist positive Borel measures $\eta_i'^{(n)}$ on $\mathbb{S}$ such that $\eta_i'^{(n)}(A_i^{(n)}) = \eta_i'^{(n)}(\mathbb{S}) = p'_{n,i}$ for $i \in I_1$, and $\sum_{i \in I_1} \eta_i'^{(n)}(A) \leq \widetilde{\mathbb{P}}_n(A)$ for all $A \in \mathcal{B}(\mathbb{S})$. Finally, let $\eta_i^{(n)} = \eta_i'^{(n)}$ for $i \in I_1 - N_n$, $\eta_{N_n}^{(n)} = \eta_{N_n}'^{(n)} + \delta_0\lambda$. Then (1.9) is obtained.

Let $\mathbb{S}' = \mathbb{S} \bigcup \{\Delta\}$, where Δ is the isolated point. We extend $\mathbb{P}_n$ from $\mathbb{S}$ to $\mathbb{S}'$ by defining $\mathbb{P}_n(\{\Delta\}) = 2^{-n}$. We consider $A_i'^{(n)} = A_i^{(n)} \bigcup \{\Delta\}$, we can easily get (1.7) with $A_i'^{(n)}$ instead of $A_i^{(n)}$ by using the same procedure.

Define $c_1^{(n)}, c_2^{(n)}, \cdots, c_{N_n}^{(n)} \in [0, 1]$ by $c_i^{(n)} = (p_{n,i} - \eta_i^{(n)}(\mathbb{S}))/p_{n,i}$. Obviously, $(1 - c_i^{(n)})\mathbb{P}_{n-1}(E_i^{(n)}) = \eta_i^{(n)}(\mathbb{S})$, $i = 1, 2, \cdots, N_n$ and

$$\mathbb{P}_{n-1}(E_0^{(n)}) + \sum_{i=1}^{N_n} c_i^{(n)} \mathbb{P}_{n-1}(E_i^{(n)}) = 1 - \sum_{i=1}^{N_n} \eta_i^{(n)}(\mathbb{S}).$$

Then, there exist $\mathbb{Q}_0^{(n)}, \mathbb{Q}_1^{(n)}, \cdots, \mathbb{Q}_{N_n}^{(n)} \in \mathcal{P}(\mathbb{S})$, such that

$$\mathbb{Q}_i^{(n)}(B)(1 - c_i^{(n)})\mathbb{P}_{n-1}(E_i^{(n)}) = \eta_i^{(n)}(B), \ i = 1, 2, \cdots, N_n, \tag{1.11}$$

$$\mathbb{Q}_0^{(n)}(B)(\mathbb{P}_{n-1}(E_0^{(n)}) + \sum_{i=1}^{N_n} c_i^{(n)} \mathbb{P}_{n-1}(E_i^{(n)})) = \mathbb{P}_n(B) - \sum_{i=1}^{N_n} \eta_i^{(n)}(B) \tag{1.12}$$

for $B \in \mathcal{B}(\mathbb{S})$.

Let $X_{n-1}, Y_0^{(n)}, \cdots, Y_{N_n}^{(n)}$ and $\xi^{(n)}$ be independent random variables on a probability space $(\Omega, \mathcal{F}, \nu)$ with $X_{n-1}, Y_0^{(n)}, \cdots, Y_{N_n}^{(n)}$ having distributions $\mathbb{P}_{n-1}, \mathbb{Q}_0^{(n)}, \cdots, \mathbb{Q}_{N_n}^{(n)}$ and $\xi^{(n)}$ having uniform distribution.

Define

$$X_n = \begin{cases} Y_i^{(n)} \text{ on } \{X_{n-1} \in E_i^{(n)}, \xi^{(n)} \geq c_i^{(n)}\}, i = 1, 2, \cdots, N_n \\ Y_0^{(n)} \text{ on } \{X_{n-1} \in E_0^{(n)}\} \bigcup \cup_{i=1}^{N_n} \{X_{n-1} \in E_i^{(n)}, \xi^{(n)} < c_i^{(n)}\}. \end{cases}$$

By (1.11) and (1.12),

$$\begin{aligned} \nu(X_n \in B) &= \sum_{i=1}^{N_n} \mathbb{Q}_i^{(n)}(B)(1 - c_i)\mathbb{P}_{n-1}(E_i^{(n)}) \\ &\quad + \mathbb{Q}_0^{(n)}(B)(\mathbb{P}_{n-1}(E_0^{(n)}) + \sum_{i=1}^{N_n} c_i^{(n)} \mathbb{P}_{n-1}(E_i^{(n)})) \\ &= \mathbb{P}_n(B), \end{aligned}$$

which implies X_n has distribution $\mathbb{P}_n$. Furthermore, since $\{X_{n-1} \in E_i^{(n)}, \xi^{(n)} > c_i^{(n)}\} \subset \{X_{n-1} \in E_i^{(n)}, X_n \in A_i^{(n)}\} \subset \{d(X_{n-1}, X_n) < 2^{-n+1}\}$,

$$\begin{aligned} \nu(d(X_{n-1}, X_n) \geq 2^{-n+1}) &\leq \nu(X_{n-1} \in E_0^{(n)}) + \nu(\bigcup_{i=1}^{N_n} \{X_{n-1} \in E_i^{(n)}, \xi^{(n)} \leq c_i^{(n)}\}) \\ &\leq \mathbb{P}_{n-1}(E_0^{(n)}) + \sum_{i=1}^{N_n} c_i^{(n)} \mathbb{P}_{n-1}(E_i^{(n)}) \\ &= \mathbb{P}_{n-1}(E_0^{(n)}) + \sum_{i=1}^{N_n} (p_{n,i} - \eta_i^{(n)}(\mathbb{S})) \leq 2^{-n+1}. \end{aligned}$$

By the Borel-Cantelli lemma, we have

$$\nu(\sum_{n=2}^{\infty} d(X_{n-1}, X_n) < \infty) = 1.$$

Thus,

$$\lim_{n \to \infty} X_n = X \ a.s..$$

Let $\mathbb{P}$ be the distribution of X, we have $\rho(\mathbb{P}_n, \mathbb{P}) \to 0$ as $n \to \infty$.

The theorem is proved. □

1.2.2 *Weak convergence and portmanteau theorem*

In this subsection, we introduce the definition and basic properties of weak convergence, the central topic in this book.

Definition 1.7. A sequence $\mathbb{P}_n \in \mathcal{P}(\mathbb{S})$ is said to converge weakly to $\mathbb{P} \in \mathcal{P}(\mathbb{S})$ if

$$\lim_{n \to \infty} \int f d\mathbb{P}_n = \int f d\mathbb{P} \tag{1.13}$$

for every $f \in C(\mathbb{S})$, denoted by $\mathbb{P}_n \Rightarrow \mathbb{P}$.

The distribution of an $\mathbb{S}-$valued random element X on $(\Omega, \mathcal{F}, \mathbb{P})$ may be denoted by $\mathbb{P}X^{-1}$. Obviously, $\mathbb{P}X^{-1} \in \mathcal{P}(\mathbb{S})$.

Definition 1.8. A sequence of $\mathbb{S}-$valued random elements X_n is said to converge weakly to $\mathbb{S}-$valued random element X if

$$\lim_{n\to\infty} \mathbb{E}f(X_n) = \mathbb{E}f(X) \tag{1.14}$$

for every $f \in C(\mathbb{S})$, denoted by $X_n \Rightarrow X$.

The random elements taking values in $\mathbb{C}[0,1]$ and $\mathbb{D}[0,1]$ are continuous path stochastic processes and the so-called càdlàg stochastic processes (which are right-continuous and have left-hand limits) respectively. Weak convergence of stochastic processes is the main topic of this book. From now on, if we do not specially mention, $\mathbb{D}[0,1]$ space is endowed with the J_1 topology and the $\mathbb{R}-$valued càdlàg process can be seen as the random element taking values in this topology space, and denote $\Delta X_t = X_t - X_{t-}$.

In fact, the weak convergence is equivalent to the convergence under Prohorov metric.

Theorem 1.2. *Let $\mathbb{P}_n \subset \mathcal{P}(\mathbb{S})$ and $\mathbb{P} \in \mathcal{P}(\mathbb{S})$, the following arguments are equivalent:*

(*i*). $\lim_{n\to\infty} \rho(\mathbb{P}_n, \mathbb{P}) = 0$;

(*ii*). $\mathbb{P}_n \Rightarrow \mathbb{P}$;

(*iii*). $\lim_{n\to\infty} \int f d\mathbb{P}_n = \int f d\mathbb{P}$ *for all uniformly continuous bounded* $f \in C(\mathbb{S})$;

(*iv*). $\limsup_{n\to\infty} \mathbb{P}_n(F) \le \mathbb{P}(F)$ *for all* $F \in \mathcal{F}(\mathbb{S})$;

(*v*). $\liminf_{n\to\infty} \mathbb{P}_n(G) \ge \mathbb{P}(G)$ *for all* $G \in \mathcal{G}(\mathbb{S})$;

(*vi*). $\lim_{n\to\infty} \mathbb{P}_n(A) = \mathbb{P}(A)$ *for all* $A \in \mathcal{B}(\mathbb{S})$ *with* $\mathbb{P}(\partial A) = 0$.

Proof.

(i)$\Rightarrow$ (ii). For each n, let $\delta_n = \rho(\mathbb{P}_n, \mathbb{P}) + \frac{1}{n}$. Consider $f \in C(\mathbb{S})$ with $f \ge 0$, we have

$$\int f d\mathbb{P}_n = \int_0^{||f||} \mathbb{P}_n(\{f > s\})ds \le \int_0^{||f||} \mathbb{P}(\{f > s\}^{\delta_n})ds + \delta_n ||f||,$$

then

$$\limsup_{n\to\infty} \int f d\mathbb{P}_n \le \lim_{n\to\infty} \int_0^{||f||} \mathbb{P}(\{f > s\}^{\delta_n})ds = \int f d\mathbb{P}. \tag{1.15}$$

Applying (1.15) to $||f|| + f$ and $||f|| - f$, (ii) is obtained.

(ii)$\Rightarrow$ (iii) is obvious.

(iii)$\Rightarrow$ (iv). For every $F \in \mathcal{F}(\mathbb{S})$, $\varepsilon > 0$, define

$$f_\varepsilon(x) = (1 - \frac{d(x, F)}{\varepsilon}) \vee 0,$$

where $d(x, F) = \inf_{y \in F} d(x, y)$, then f_ε is uniform continuous. By (iii),

$$\limsup_{n\to\infty} \mathbb{P}_n(F) \le \lim_{n\to\infty} \int f_\varepsilon d\mathbb{P}_n = \int f_\varepsilon d\mathbb{P} \to \mathbb{P}(F)$$

as $\varepsilon \downarrow 0$.

(iv)$\Rightarrow$ (v). For every $G \in \mathcal{G}(\mathbb{S})$,

$$\liminf_{n\to\infty} \mathbb{P}_n(G) \ge 1 - \limsup_{n\to\infty} \mathbb{P}_n(G^c) \ge 1 - \mathbb{P}(G^c) = \mathbb{P}(G)$$

by (iv).

(v)$\Rightarrow$ (vi). Let $A \in \mathcal{B}(\mathbb{S})$ with $\mathbb{P}(\partial A) = 0$.

$$\limsup_{n\to\infty} \mathbb{P}_n(A) \le \limsup_{n\to\infty} \mathbb{P}_n(\overline{A}) = 1 - \liminf_{n\to\infty} \mathbb{P}_n(\overline{A}^c) \le 1 - \mathbb{P}(\overline{A}^c) = \mathbb{P}(A)$$

and

$$\liminf_{n\to\infty} \mathbb{P}_n(A) \ge \lim_{n\to\infty} \mathbb{P}_n(A - \partial A) \ge \mathbb{P}(A - \partial A) = \mathbb{P}(A),$$

which imply (vi).

(vi)$\Rightarrow$ (ii). Let $f \in C(\mathbb{S})$ and $f \ge 0$. Since $\partial\{f \ge t\} \subset \{f = t\}$, and $\mathbb{P}(\{f = t\}) = 0$ for all but at most countably many $t \ge 0$,

$$\lim_{n\to\infty} \int f d\mathbb{P}_n = \lim_{n\to\infty} \int_0^{||f||} \mathbb{P}_n(\{f \ge t\})dt = \int_0^{||f||} \mathbb{P}(\{f \ge t\})dt = \int f d\mathbb{P}.$$

(v)$\Rightarrow$ (i). For any $\varepsilon > 0$, let $E_1, E_2, \cdots \in \mathcal{B}(\mathbb{S})$ be a partition of $\mathbb{S}$ with diameters less than $\varepsilon/2$. There exists n_0, for $n \ge n_0$,

$$\mathbb{P}(\bigcup_{i=1}^{n} E_i) \ge 1 - \varepsilon/2.$$

Let $\mathfrak{G}$ be the finite collection of the sets of the form $(\bigcup_{i\in I} E_i)^{\varepsilon/2}$, where $I \subset \{1, 2, \cdots, n_0\}$. Then there exists n_1, for every $n \ge n_1$,

$$\mathbb{P}(G) \le \mathbb{P}_n(G) + \varepsilon/2$$

for every $G \in \mathfrak{G}$ by (v).

For any $F \in \mathcal{F}(\mathbb{S})$, let

$$F_0 = \bigcup_{i=1}^{n_0} \{E_i : E_i \cap F \ne \emptyset\}.$$

We have

$$\mathbb{P}(F) \le \mathbb{P}(F_0^{\varepsilon/2}) + \varepsilon/2 \le \mathbb{P}_n(F_0^{\varepsilon/2}) + \varepsilon \le \mathbb{P}_n(F^\varepsilon) + \varepsilon,$$

hence for $n \ge n_1$, $\rho(\mathbb{P}_n, \mathbb{P}) \le \varepsilon$. (i) is proved. □

Proposition 1.7. *If $X_n, Y_n, n \ge 1$, X are random elements which take values in $\mathbb{S}$, and $X_n \Rightarrow X$, $d(X_n, Y_n) \to 0$ as $n \to \infty$, then $Y_n \Rightarrow X$.*

Proof.
Let $F \in \mathcal{F}(\mathbb{S})$, $F_\varepsilon = \{x : d(x, F) \le \varepsilon\}$.

$$\mathbb{P}(Y_n \in F) \le \mathbb{P}(d(X_n, Y_n) \ge \varepsilon) + \mathbb{P}(X_n \in F_\varepsilon).$$

Hence,

$$\limsup_{n\to\infty} \mathbb{P}(Y_n \in F) \le \limsup_{n\to\infty} \mathbb{P}(X_n \in F_\varepsilon) \le \mathbb{P}(X \in F_\varepsilon) \to \mathbb{P}(X \in F)$$

as $\varepsilon \downarrow 0$. □

Proposition 1.8. *Suppose h is a measurable mapping from $\mathbb{S}$ to $\mathbb{S}'$, and let D_h be the set of discontinuous points of h. If $\mathbb{P}_n \Rightarrow \mathbb{P} \in \mathcal{P}(\mathbb{S})$ and $\mathbb{P}(D_h) = 0$, then $\mathbb{P}_n h^{-1} \Rightarrow \mathbb{P}h^{-1}$.*

Proof. Let $F \in \mathcal{F}(\mathbb{S}')$. Since $\mathbb{P}(D_h) = 0$ and $\overline{h^{-1}F} \subset D_h \bigcup (h^{-1}F)$,

$$\limsup_{n\to\infty} \mathbb{P}_n h^{-1}(F)) \le \limsup_{n\to\infty} \mathbb{P}_n(\overline{h^{-1}F}) \le \mathbb{P}(\overline{h^{-1}F}) = \mathbb{P}h^{-1}(F).$$ □

Proposition 1.8 is an extension of the continuous mapping theorem. The following propositions provide some examples of the continuous mapping theorem.

Proposition 1.9. *Suppose $X^n, n \ge 1$, X are càdlàg stochastic processes. If $X^n \Rightarrow X$ and $\Delta X_1 = 0$ a.s., then*

$$\sup_{0\le t\le 1} |X_t^n| \Rightarrow \sup_{0\le t\le 1} |X_t|,$$

$$\sup_{0\le t\le 1} |\Delta X_t^n| \Rightarrow \sup_{0\le t\le 1} |\Delta X_t|.$$

Proof. By Proposition 1.8, we only need to prove that

$$\lim_{n\to\infty} \sup_{0\le t\le 1} |\alpha_n(t)| = \sup_{0\le t\le 1} |\alpha(t)|, \tag{1.16}$$

$$\lim_{n\to\infty} \sup_{0\le t\le 1} |\Delta\alpha_n(t)| = \sup_{0\le t\le 1} |\Delta\alpha(t)|, \tag{1.17}$$

if $\alpha_n, \alpha \in \mathbb{D}[0,1]$, $\Delta\alpha(1) = 0$ and $\alpha_n \xrightarrow{J_1} \alpha$ as $n \to \infty$.

Since $\alpha_n \xrightarrow{J_1} \alpha$ as $n \to \infty$, there exists a continuous one-to-one mapping sequence $\lambda_n(t)$ of $[0,1]$ onto $[0,1]$ such that

$$\lim_{n\to\infty} \sup_{t\in[0,1]} |\alpha_n(\lambda_n(t)) - \alpha(t)| = 0, \qquad \lim_{n\to\infty} \sup_{t\in[0,1]} |\lambda_n(t) - t| = 0.$$

We have

$$\begin{aligned}
&\Big| \sup_{0\le t\le 1} |\alpha_n(t)| - \sup_{0\le t\le \lambda_n^{-1}(1)} |\alpha(t)| \Big| \\
&= \Big| \sup_{0\le t\le \lambda_n^{-1}(1)} |\alpha_n(\lambda_n(t))| - \sup_{0\le t\le \lambda_n^{-1}(1)} |\alpha(t)| \Big| \\
&\le \sup_{0\le t\le \lambda_n^{-1}(1)} |\alpha_n(\lambda_n(t)) - \alpha(t)| \to 0.
\end{aligned}$$

(1.16) is showed, since $\lambda_n^{-1}(1) \to 1$ as $n \to \infty$ and $\Delta\alpha(1) = 0$. Similarly, (1.17) holds true. □

Proposition 1.10. *Suppose $X^n, n \geq 1$, X are càdlàg stochastic processes. g is a continuous function: $\mathbb{R} \to \mathbb{R}$, and vanishing at the neighborhood of 0. If $X^n \Rightarrow X$, then*

$$(X^n, \sum_{s \leq \cdot} g(\Delta X_s^n)) \Rightarrow (X, \sum_{s \leq \cdot} g(\Delta X_s)).$$

The proof of this proposition is complex, and can be founded in *Jacod and Shiryaev (2003).*

1.3 How to verify the weak convergence?

When we intend to prove the weak convergence of probability measures and stochastic processes, firstly, it is needed to prove the relative compactness of $\{\mathbb{P}_n\}$, the probability measure sequence or distribution sequence of stochastic processes. Relative compactness means that for $\{\mathbb{P}_{n_k}\}$, subsequence of $\{\mathbb{P}_n\}$, there exists a further subsequence $\{\mathbb{P}_{n_{k_j}}\}$, such that $\mathbb{P}_{n_{k_j}}$ converges. Furthermore, it is needed to identify the limit. If the limiting measure of weak convergence of $\mathbb{P}_{n_{k_j}}$ is common for every subsequence $\mathbb{P}_{n_k}$, we can identify the limit of $\{\mathbb{P}_n\}$.

In the previous section, we have already studied the topology of the $\mathcal{P}(\mathbb{S})$, and prove that the Prohorov topology is equivalent to the weak topology, which is generated by the weak convergence. Consequently, a characterization of compact subset of $\{\mathbb{P}_n\}$ under the Prohorov topology is crucial for the relative compactness of $\{\mathbb{P}_n\}$. In this section, we will characterize the relative compactness of the subsets of $\mathcal{P}(\mathbb{S})$ by the notion of *tightness.*

1.3.1 *Tightness*

Definition 1.9. A subset $\mathcal{M} \subset \mathcal{P}(\mathbb{S})$ is said to be *tight*, if for every $\varepsilon > 0$, there exists a compact set $K \in \mathbb{S}$, such that $\mathbb{P}(K) \geq 1 - \varepsilon$ for every $\mathbb{P} \in \mathcal{M}$.

Theorem 1.3. *(Prohorov Theorem) Let $\mathcal{K} \subset \mathcal{P}(\mathbb{S})$, then the followings are equivalent:*

(i) $\mathcal{K}$ is relative compactness.

(ii) $\mathcal{K}$ is tight.

Proof. Since $(\mathbb{S}, \rho)$ is completely separable, the relative compactness of $\mathcal{K}$ is equivalent to the totally bounded.

(i)$\Rightarrow$ (ii). Let $\varepsilon > 0$, since $\mathcal{K}$ is totally bounded, there exists a finite subset $\mathcal{T}_n \subset \mathcal{K}$ such that $\mathcal{K} \subset \{\mathbb{Q} : \rho(\mathbb{Q}, \mathbb{P}) < \frac{\varepsilon}{2^{n+1}}$ for some $\mathbb{P} \in \mathcal{T}_n\}$ for every $n = 1, 2, \cdots$.

For one measure $\mathbb{P}$ in $\mathcal{P}(\mathbb{S})$, the tightness is obvious for $(\mathbb{S}, d)$, which is complete and separable. In fact, for every $\varepsilon > 0$, there exists $G_n \in \mathcal{G}(\mathbb{S})$, such that $\mathbb{P}(G_n) \geq 1 - \frac{\varepsilon}{2^n}$. Let K be the closure of $\bigcap_{n=1}^{\infty} G_n$, then $\mathbb{P}(K) \geq 1 - \sum_{n=1}^{\infty} \frac{\varepsilon}{2^n} = 1 - \varepsilon$.

Then for $\mathcal{T}_n$, there exists K_n such that $\mathbb{P}(K_n) \geq 1 - \frac{\varepsilon}{2^{n+1}}$ for every $\mathbb{P} \in \mathcal{T}_n$ and $\varepsilon > 0$. Given $\mathbb{Q} \in \mathcal{K}$, it means that for $n = 1, 2 \cdots$, there exists $\mathbb{P}_n \in \mathcal{T}_n$, such that

$$\mathbb{Q}(K_n^{\varepsilon/2^{n+1}}) \geq \mathbb{P}_n(K_n) - \frac{\varepsilon}{2^{n+1}} \geq 1 - \frac{\varepsilon}{2^n}.$$

Let K be the closure of $\bigcap_{n=1}^{\infty} K_n^{\varepsilon/2^{n+1}}$, then $\mathbb{Q}(K) \geq 1 - \varepsilon$.

(ii)$\Rightarrow$ (i). We only need to construct that for given δ, there exists a finite subset $\mathcal{N}$ of $\mathcal{P}(\mathbb{S})$, such that $\mathcal{K} \subset \{\mathbb{Q} : \rho(\mathbb{P}, \mathbb{Q}) < \delta$ for some $\mathbb{P} \in \mathcal{N}\}$.

For $0 < \varepsilon < \delta/2$, there exists a compact set K, such that $\mathbb{P}(K) \geq 1 - \varepsilon$ for every $\mathbb{P} \in \mathcal{K}$. By the compactness of K, there exists a subset $\{x_1, x_2, \cdots, x_k\} \subset K$, such that $K^{\varepsilon} \subset \bigcup_{i=1}^{k} B_i$, where B_i is the open ε neighborhood of x_i.

We construct $\mathcal{N}$ as follows: given $x_0 \in \mathcal{S}$, $n \geq k/\varepsilon$, $\mathcal{N}$ is the collection of the probability measures of the form

$$\mathbb{P} = \sum_{i=0}^{k} (\frac{j_i}{n}) \delta_{x_i},$$

where $0 \leq j_i \leq n$, and $\sum_{i=0}^{k} j_i = n$.

For fixed $\mathbb{Q} \in \mathcal{K}$, let $j_i = [n\mathbb{Q}(A_i)]$, where $A_i = B_i - \bigcup_{l=1}^{i-1} B_l$, $j_0 = n - \sum_{i=1}^{k} j_i$. For every $F \in \mathcal{F}(\mathbb{S})$, we have

$$\begin{aligned}\mathbb{Q}(F) &\leq \mathbb{Q}(\bigcup_{F \cap A_i \neq \emptyset} A_i) + \varepsilon \\ &\leq \sum_{F \cap A_i \neq \emptyset} \frac{[n\mathbb{Q}(A_i)]}{n} + \frac{k}{n} + \varepsilon \\ &\leq \mathbb{P}(F^{2\varepsilon}) + 2\varepsilon.\end{aligned}$$

$\rho(\mathbb{P}, \mathbb{Q}) \leq \varepsilon$ is obtained. □

For the weak convergence of stochastic processes, the tightness of the distribution sequence of stochastic processes is really very abstractive. We will give some criterions for the tightness of distributions of càdlàg processes.

Let $\alpha \in \mathbb{D}[0, 1]$,

$$\omega'(\alpha, \theta) = \inf_{r>0} \{\max_{i \leq r} \sup_{t_{i-1} \leq s,t < t_i} |\alpha(s) - \alpha(t)| : 0 = t_0 < \cdots < t_r = 1, \inf_{i \leq r}(t_i - t_{i-1}) \geq \theta\}.$$

Lemma 1.1. *(Skorohod (1956a)) A subset A of $\mathbb{D}[0, 1]$ is relatively compact for J_1 topology if and only if*

$$\begin{cases} \sup_{\alpha \in A} \sup_{s \in [0,1]} |\alpha(s)| < \infty, \\ \lim_{\theta \downarrow 0} \sup_{\alpha \in A} \omega'(\alpha, \theta) = 0. \end{cases} \tag{1.18}$$

Theorem 1.4. *The càdlàg process sequence $\{X^n\}$ is tight if and only if*

(i) for every $\varepsilon > 0$, there exist positive integer n_0 and $K \in \mathbb{R}_+$, such that

$$\mathbb{P}(\sup_{t \in [0,1]} |X_t^n| > K) \leq \varepsilon$$

for every $n \geq n_0$.

(ii) for every $\varepsilon > 0$, $\delta > 0$, *there exist positive integer* n_0 *and* $\theta > 0$, *such that*

$$\mathbb{P}(\omega'(X^n, \theta) \geq \delta) \leq \varepsilon$$

for every $n \geq n_0$.

Proof. Sufficiency. We suppose that (i) and (ii) hold. Fix $\varepsilon > 0$ and positive integer k, there exist K_ε and θ_k, such that

$$\mathbb{P}(\sup_{t\in[0,1]} |X_t^n| > K_\varepsilon) \leq \frac{\varepsilon}{2}, \ \mathbb{P}(\omega'(X^n, \theta_k) \geq 1/k) \leq \frac{\varepsilon}{2^{k+1}}$$

for every $n \geq n_0$.

Let $A_\varepsilon = \{\alpha \in \mathbb{D}[0,1] : \sup_{t\in[0,1]} |\alpha(t)| \leq K_\varepsilon, \omega'(\alpha, \theta_k) \leq 1/k$ for all $k \in \mathbb{N}\}$. Obviously,

$$\mathbb{P}(X^n \overline{\in} A_\varepsilon) \leq \mathbb{P}(\sup_{t\in[0,1]} |X_t^n| > K_\varepsilon) + \sum_{k\geq 1} \mathbb{P}(\omega'(X^n, \theta_k) \geq 1/k) \leq \varepsilon,$$

and A_ε is relatively compact in $\mathbb{D}[0,1]$ by Lemma 1.1. This implies tightness of $\{X^n\}$.

Necessity. By Theorem 1.3, for every $\varepsilon > 0$, there exists compact set $K \in \mathbb{D}[0,1]$, such that $\mathbb{P}(X^n \in K) \geq 1 - \varepsilon$ for all n. Then for every $\delta > 0$, there exists $\theta > 0$ with $\sup_{\alpha\in K} \omega'(\alpha, \theta) \leq \delta$, and $T := \sup_{\alpha\in K, t\in[0,1]} |\alpha(t)|$ is finite by Lemma 1.1. Thus,

$$\mathbb{P}(\sup_{t\in[0,1]} |X_t^n| > T) \leq \varepsilon, \ \mathbb{P}(\omega'(X^n, \theta) \geq \delta) \leq \varepsilon$$

for all n. These imply (i) and (ii) with $n_0 = 1$ □

In the study of weak convergence for sums of some stochastic processes, the following theorem is very interesting.

Theorem 1.5. *Suppose that for all* $n, p \in \mathbb{N}$, *the càdlàg process sequence* $\{X^n\}$ *has a decomposition*

$$X^n = Y^n + Z^{np} + W^{np},$$

such that

(i) the sequence $\{Y^n\}_{n\geq 1}$ *is tight;*

(ii) the sequence $\{Z^{np}\}_{n\geq 1}$ *is tight and there is a sequence* $\{a_p\}$ *with* $\lim_{p\to\infty} a_p = 0$, *and*

$$\lim_{n\to\infty} \mathbb{P}(\sup_{t\in[0,1]} |\Delta Z_t^{np}| \geq a_p) = 0;$$

(iii) for any $\varepsilon > 0$,

$$\lim_{p\to\infty} \limsup_{n\to\infty} \mathbb{P}(\sup_{t\in[0,1]} |W_t^{np}| \geq \varepsilon) = 0.$$

Then $\{X^n\}$ *is tight.*

Proof. It is obvious that condition (i) in Theorem 1.4 is satisfied.

Let $\varepsilon, \eta > 0$, there exist p_0, n_0 and $\theta > 0$, such that $a_p \leq \eta$ for $p \geq p_0$,

$$\mathbb{P}(\omega'(Y^n, \theta) \geq \eta) \leq \varepsilon, \ \mathbb{P}(\omega'(Z^{np}, 2\theta) \geq \eta) \leq \varepsilon,$$

$$\mathbb{P}(\sup_{t\in[0,1]} |W_t^{np}| \geq \eta) \leq \varepsilon, \ \mathbb{P}(\sup_{t\in[0,1]} |\Delta Z_t^{np}| \geq \eta) \leq \varepsilon$$

for $n \geq n_0$. Thus

$$\mathbb{P}(\omega'(X^n, \theta) > 6\eta) \leq 6\varepsilon,$$

since

$$\begin{aligned}\omega'(X^n, \theta) &\leq \omega'(Y^n + Z^{np}, \theta) + \sup\{\sup_{s\leq t,u\leq s+\theta} |W_t^{np} - W_u^{np}| : 0 \leq s \leq s+\theta \leq 1\} \\ &\leq \omega'(Y^n, \theta) + 2\omega'(Z^{np}, \theta) + \sup_{t\in[0,1]} |\Delta Z_t^{np}| + 2 \sup_{t\in[0,1]} |W_t^{np}|.\end{aligned}$$

(ii) in Theorem 1.4 is satisfied. □

1.3.2 *Identifying the limit*

Suppose that $\{\mathbb{P}_n\}$ is relatively compact. Then, each subsequence contains a further subsequence converging weakly to some limit $\mathbb{Q}$. If every such limit coincides with each other, we can obtain $\{\mathbb{P}_n\}$ weakly converge to $\mathbb{Q}$ as $n \to \infty$. This procedure is called as identifying the limit.

Let $X^n = (X_t^n)_{t\in[0,1]}$ and $X = (X_t)_{t\in[0,1]}$ be stochastic processes defined on the probability space $(\Omega, \mathcal{F}, \mathbb{P})$. Let $\pi_{t_1\cdots t_k}$ be the mapping that carries the point $x \in \mathbb{C}[0,1]$ or $\mathbb{D}[0,1]$ to the point $(x(t_1), \cdots, x(t_k))$ of $\mathbb{R}^k$.

Proposition 1.11. *Suppose that $X^n = (X_t^n)_{t\in[0,1]}$, $n \geq 1$, are continuous path processes on $(\Omega, \mathcal{F}, \mathbb{P})$ with tightness. If the finite dimensional distributions of $\{X^n\}$ weakly converge, then $\{X^n\}$ weakly converge.*

Proof. We assume that $\{\mathbb{P}(X_{t_1,\cdots,t_k}^n)^{-1}\}$ weakly converge to $\{\mathbb{P}X_{t_1,\cdots,t_k}^{-1}\}$ for every $t_1, t_2, \cdots, t_k \in [0,1]$. Tightness implies each subsequence $\{X^{n'}\}$ contains a further subsequence $\{X^{n''}\}$ converging weakly to some limit X'. If we can prove that $\{\mathbb{P}X_{t_1,\cdots,t_k}^{-1}\} = \{\mathbb{P}X_{t_1,\cdots,t_k}'^{-1}\}$, we will obtain this proposition. It is enough to prove that the finite-dimensional sets form a determining class of $\mathbb{C}[0,1]$ (c.f., page 19 *Billingsley (1999)*). In fact, when $x \in \mathbb{C}[0,1]$, $\pi_{t_1\cdots t_k}$ is obviously a continuous mapping. Define *the finite dimensional sets* as the sets of the form $\pi_{t_1\cdots t_k}^{-1}H$ with $H \in \mathcal{B}(\mathbb{R}^k)$. $\pi_{t_1\cdots t_k}^{-1}H \in \mathcal{B}(\mathbb{C}[0,1])$, since $\pi_{t_1\cdots t_k}$ is continuous. The closed sphere $\{y : \delta(x,y) \leq \varepsilon\}$ is the limit of the finite dimensional sets $\{y : |x(\frac{i}{n}) - y(\frac{i}{n})| \leq \varepsilon, \ i = 1, 2, \cdots, n\}$. Since $\mathbb{C}[0,1]$ is separable and complete, the finite dimensional sets generate $\mathcal{B}(\mathbb{C}[0,1])$. □

However, π_t is not always continuous in $\mathbb{D}[0,1]$. The following lemma is about this fact.

Lemma 1.2. *For $x \in \mathbb{D}[0,1]$, π_t is continuous at x if and only if x is continuous at t.*

Proof. If $x_n \xrightarrow{J_1} x$, and x is continuous at t, then there exists λ_n such that

$$|x_n(t) - x(t)| \le |x_n(t) - x(\lambda_n^{-1}(t))| + |x(\lambda_n^{-1}(t)) - x(t)| \to 0$$

as $n \to \infty$, i.e., π_t is continuous at x.

On the other hand, suppose that x is discontinuous at t. Let $\lambda_n(t) = t - 1/n$, and λ_n be linear on $[0,t]$ and on $[t,1]$. Let $x_n = x(\lambda_n(t))$, then $x_n \xrightarrow{J_1} x$, but $x_n(t)$ does not converge to $x(t)$. □

We assume that X is a càdlàg process on $[0,1]$, and for every t, $\mathbb{P}(\Delta X_t \neq 0) = 0$. X is called as the process with not fixed discontinuous point. From Proposition 1.11, we can obtain the following proposition.

Proposition 1.12. *Suppose that $\{X^n = (X_t^n)_{t\in[0,1]}\}$ are càdlàg processes on $(\Omega, \mathcal{F}, \mathbb{P})$, and $\{X^n\}$ is tight. If the finite dimensional distributions of $\{X^n\}$ weakly converge to that of X. Then $\{X^n\}$ weakly converge to X.*

The processes with not fixed discontinuous point contain a lot of important examples of stochastic processes, such as Lévy processes.

In the previous two propositions, we have introduced important rules of identifying the limiting process. Usually, when the limiting process is Lévy process, people can compute the finite dimensional distributions of the underlying process. Then convergence of finite dimensional distributions is the tool for identifying the limit.

However, when the limiting process is more general diffusion process, its finite dimensional distributions are difficult to compute. They can be determined by the generator of the diffusion process, and furthermore, the convergence of finite dimensional distributions is equivalent to convergence of martingale characteristics for generator. Thus convergence of martingale characteristics for the generator can identify the limiting processes (c.f., page 226 *Ethier and Kurtz (1986)*). Moreover, when the limiting process is semimartingale, its predictable characteristics play the similar roles as martingale characteristics for generator. More details can be found in Chapter 3 of this book.

1.4 Two examples of applications of weak convergence

In this section, we will present two examples of applications of weak convergence.

1.4.1 *Unit root testing*

Methods for detecting the presence of a unit root in parametric time series models have attracted a good deal of interest. A field of applications where the hypothesis of a unit root has important implications is econometrics. This is because a unit root is often a theoretical implication of models which postulate the rational use of information that is available to economic agents.

Let

$$Y_i = \alpha Y_{i-1} + X_i,$$

where $\{X_i\}$ is a mean zero stationary time series. The unit root testing is aim to test whether $\alpha = 1$ from $\{Y_i\}$, and it can also be used for testing stationarity for $\{Y_i\}$.

Usually, we use statistic

$$\hat{\alpha} = \frac{\sum_{i=1}^{n} Y_{i-1} Y_i}{\sum_{i=1}^{n} Y_{i-1}^2}$$

for unit root testing.

Theorem 1.6. *Assume* $\frac{1}{\sqrt{n}} \sum_{i=1}^{[n\cdot]} X_i \Rightarrow \sigma^2 W(\cdot)$, *where* W *is a standard Brownian motion. We have*

$$n(\hat{\alpha} - 1) \xrightarrow{d} \frac{\lambda + \sigma^2 \int_0^1 W(v) dW(v)}{\sigma^2 \int_0^1 W^2(v) dv}.$$

The proof of this theorem will be given in Chapter 3.

1.4.2 *Goodness-of-fit testing for volatility*

In general, the diffusion process $(X_t)_{t>0}$ is a solution of the following stochastic differential equation:

$$dX_t = b(t, X_t)dt + \sigma(t, X_t)dW_t$$

where $(W_t)_{t>0}$ is a standard Brownian motion. The specification of a parametric form for the volatility $\sigma(t, X_t)$ is quite an interesting statistical problem.

Consider the null hypothesis:

$$\mathrm{H}_0 : \sigma(s, X_s) = \sum_{j=1}^{d} \theta_j \sigma_j(s, X_s),$$

where $\theta = (\theta_1, \theta_2, \cdots, \theta_d)$ are unknown parameters and $\sigma_1, \sigma_2, \cdots, \sigma_d$ are given and known volatility functions.

For testing the null hypothesis, we consider the following stochastic process

$$M_t := \int_0^t \Big\{\sigma(s, X_s) - \sum_{j=1}^{d} \theta_j^{min} \sigma_j(s, X_s)\Big\} ds$$

where the vector $\theta^{min} = (\theta_1^{min}, \cdots, \theta_d^{min})$ is defined by

$$\theta^{min} := \arg\min_{\theta \in \mathcal{R}^d} \int_0^1 \Big\{ \sigma(s, X_s) - \sum_{j=1}^{d} \theta_j \sigma_j(s, X_s) \Big\} ds.$$

Based on the estimations of certain integrals of the volatility, we can construct the the empirical value of M_t, $\hat{M}_t$ (c.f. *Dette and Podolskij (2008)*).

Dette and Podolskij (2008) have proved

$$A_n := \sqrt{n}\hat{M} \Rightarrow A$$

on $D[0,1]$, where A is a stochastic integral. Based on weak convergence, in the case of the Kolmogorov-Smirnov statistics

$$\sqrt{n} \sup_{t \in [0,1]} |\hat{M}_t| \xrightarrow{d} \sup_{t \in [0,1]} |A_t|$$

by Proposition 1.8. Moreover, consider local alternatives of the form

$$H_1^{(n)} : \sigma(t, X_t) = \sum_{j=1}^{d} \theta_j \sigma_j(s, X_s) + n^{-1/2} h(t, X_t).$$

Under $H_1^{(n)}$, Dette and Podolskij (2008) shows

$$\sqrt{n}\hat{M} \Rightarrow \widetilde{A} \tag{1.19}$$

on $D[0,1]$, where

$$\widetilde{A}_t = A_t + \Big(\int_0^t h(s, X_s) ds - t \int_0^1 h(s, X_s) ds \Big).$$

The goodness-of-fit test based on (1.19) is more powerful with respect to the Pitman alternatives than the test which rejects the null hypothesis of homoscedasticity for large values of the statistic $\hat{M}_t$.

Chapter 2

Convergence to the Independent Increment Processes

In the previous chapter, we introduce the general criterion for the weak convergence of stochastic processes. When the limiting process is independent increment process, the sufficient conditions of weak convergence can be weakened. In this chapter, we will discuss these in details.

2.1 The basic conditions of convergence to the Gaussian independent increment processes

Brownian motion is a very important example of independent increment processes, it is the limiting process in a lot of weak convergence theorems. Thus it is very interesting to study the convergence to Brownian motion. For simplicity, we firstly study weak convergence of random elements to Brownian motion in the space $\mathbb{C}[0,1]$, then extend the result to $\mathbb{D}[0,1]$.

In 1.1.1, we define the local uniform topology in $\mathbb{C}[0,1]$

$$\rho(\alpha,\beta) = \sup_{s\in[0,1]} |\alpha(s) - \beta(s)|$$

and give the sufficient and necessary conditions for tightness in $(\mathbb{C}[0,1], \mathcal{B}(\mathbb{C}[0,1]))$.

Let $\mathbb{P}_n$ be probability measures on $(\mathbb{C}[0,1], \mathcal{B}(\mathbb{C}[0,1]))$. We have:

Theorem 2.1. *$\{\mathbb{P}_n\}_{n\geq 1}$ is tight if and only if (iff)*

(i) for every $\varepsilon > 0$, there exist $a > 0$ and n_0, such that

$$\mathbb{P}_n\{x : |x(0)| \geq a\} \leq \varepsilon$$

as $n \geq n_0$;

(ii) for every $\varepsilon > 0$, $\eta > 0$, there exist δ, $0 < \delta < 1$, and n_0, such that

$$\mathbb{P}_n\{x : \sup_{|s-t|\leq\delta} |x(s) - x(t)| \geq \eta\} \leq \varepsilon$$

as $n \geq n_0$.

Proof. Sufficiency. Since $(\mathbb{C}[0,1], \mathcal{B}(\mathbb{C}[0,1]))$ is separable and complete, for fixed n, $\{\mathbb{P}_i : i \leq n\}$ is obviously tight. For convenience, we may assume that $n_0 = 1$.

From (i) and (ii), for given $\varepsilon > 0$ and integer $k > 0$, there exist $a > 0$ and $\delta_k > 0$ such that

$$\mathbb{P}_n\{x : |x(0)| \leq a\} \geq 1 - \varepsilon,$$

$$\mathbb{P}_n\{x : \sup_{|s-t|\leq\delta_k} |x(s) - x(t)| \leq \frac{1}{k}\} \geq 1 - \frac{\varepsilon}{2^k}$$

for all n. Let K be the closure of $\{x : |x(0)| \leq a\} \bigcap \{x : \sup_{|s-t|\leq\delta_k} |x(s)-x(t)| \leq \frac{1}{k}\}$, then $\mathbb{P}_n(K) \geq 1 - 2\varepsilon$, so $\{\mathbb{P}_n\}$ is tight.

Necessity. Suppose that $\{\mathbb{P}_n\}$ is tight. For given ε, we have compact subset K such that $\mathbb{P}_n(K) \geq 1 - \varepsilon$ for all n. By the Arzalà-Ascoli theorem, we have $K \subset \{x : |x(0)| \leq a\}$ for a large enough and $K \subset \{x : \sup_{|s-t|\leq\delta} |x(s) - x(t)| \geq \eta\}$ for η small enough. (i) and (ii) are proved. □

In some situations, for example, in some statistical applications we need to study the weak convergence of random elements which take values in $\mathbb{D}[0,1]$. First of all, we discuss the tightness of probability measures on $(\mathbb{D}[0,1], \mathcal{B}(\mathbb{D}[0,1]))$.

For $\alpha \in \mathbb{D}[0,1]$, recall

$$\omega'(\alpha, \delta) = \inf_{r>0}\{\max_{i\leq r} \sup_{s,t\in[t_{i-1},t_i)} |\alpha(s)-\alpha(t)| : 0 = t_0 < \cdots < t_r = 1, \ \inf_{i\leq r}(t_i-t_{i-1}) \geq \delta\}.$$

Let $\mathbb{P}_n$ be probability measures on $(\mathbb{D}[0,1], \mathcal{B}(\mathbb{D}[0,1]))$. We have:

Theorem 2.2. *$\{\mathbb{P}_n\}_{n\geq1}$ is tight iff*

(i) for every $\varepsilon > 0$, there exist $a > 0$ and n_0, such that

$$\mathbb{P}_n\{x : \sup_{t\in[0,1]} |x(t)| \geq a\} \leq \varepsilon$$

as $n \geq n_0$;

(ii) for every $\varepsilon > 0$, $\eta > 0$, there exist δ, $0 < \delta < 1$, and n_0, such that

$$\mathbb{P}_n\{x : \omega'(x, \delta) \geq \eta\} \leq \varepsilon$$

as $n \geq n_0$.

Proof. Sufficiency. For fixed n, $\{\mathbb{P}_i : i \leq n\}$ is obviously tight. For convenience, we may assume that $n_0 = 1$.

Fix $\varepsilon > 0$ and integer $k > 0$, let $a_\varepsilon < \infty$ and $\delta_{\varepsilon k} > 0$ satisfy

$$\sup_{n\geq1} \mathbb{P}_n\{x : \sup_{t\in[0,1]} |x(t)| \geq a_\varepsilon\} \leq \frac{\varepsilon}{2},$$

$$\sup_{n\geq1} \mathbb{P}_n\{x : \omega'(x, \delta_{\varepsilon k}) \geq \frac{1}{k}\} \leq \frac{\varepsilon}{2^{k+1}}.$$

Let A_ε be the closure of

$$\{x : \sup_{t\in[0,1]} |x(t)| \leq a_\varepsilon, \omega'(x, \delta_{\varepsilon k}) \leq \frac{1}{k} \text{ for } k \geq 1\},$$

then

$$\mathbb{P}_n\{x : x \in A_\varepsilon^c\} \leq \mathbb{P}_n\{x : \sup_{t\in[0,1]} |x(t)| \geq a_\varepsilon\} + \sum_{k=1}^{\infty} \mathbb{P}_n\{x : \omega'(x, \delta_{\varepsilon k}) \geq \frac{1}{k}\} < \varepsilon.$$

Hence $\{\mathbb{P}_n\}$ is tight.

Necessity. Suppose that $\{\mathbb{P}_n\}$ is tight. For given ε, we have compact subset K such that $\mathbb{P}_n(K) \geq 1 - \varepsilon$ for all n. By the Arzalà-Ascoli theorem, we have $K \subset \{x : \sup_{t\in[0,1]} |x(t)| \leq a\}$ for a, large enough and $K \subset \{x : \omega'(x, \delta) \geq \eta\}$ for η, small enough. (i) and (ii) are proved. □

Let us come back to the field of weak convergence of stochastic processes. When the limiting processes are Gaussian independent and stationary increment processes, which are processes with not fixed discontinuous points, the convergence of finite dimensional distributions can identify the limiting processes by Proposition 1.12.

2.2 Donsker invariance principle

In the classical probability theory, the central limit theorem is a fundamental theorem. It tells us that, quite generally, what happens when we have the sum of a large number of random variables each of which contributes a small amount to the total. In fact, the central limit theorem provides a universal property in classical probability theory.

The central limit theorem describes weak convergence of probability measure on $\mathbb{R}$. In the other words, central limit theorem depicts weak convergence of $\mathbb{R}-$valued random elements. It is natural to extend central limit theorem to more general cases. *Donsker invariance principle* is an extension of cental limit theorem on random elements to $\mathbb{C}[0, 1]$.

2.2.1 *Classical Donsker invariance principle*

Consider a sequence $\{X_i\}_{i\geq 1}$ of independent and identically distributed (i.i.d.) random variables with $\mathbb{E}X_i = 0$ and $\mathbb{E}X_i^2 = \sigma^2 < \infty$. Let us consider a continuous partial sum process on $[0, 1]$:

$$W_t^n = \frac{1}{\sqrt{n}\sigma} \sum_{i=1}^{[nt]} X_i + (nt - [nt]) \frac{X_{[nt]+1}}{\sqrt{n}\sigma} \tag{2.1}$$

where $[nt]$ is an integer part of nt. W^n is a random element in $\mathbb{C}[0, 1]$.

Theorem 2.3. *If $\{X_i\}_{i\geq 1}$ are* i.i.d. *random variables with mean 0 and variance σ^2. W_t^n is defined as (2.1). Then $W^n \Rightarrow W$, where W is a Brownian motion.*

Proof. By the previous section, we will obtain this theorem through showing

$$(W_{t_1}^n, \cdots, W_{t_k}^n) \xrightarrow{d} (W_{t_1}, \cdots, W_{t_k}) \tag{2.2}$$

for every $t_1, \cdots, t_k \in [0,1]$ and

$$\lim_{\delta\to 0}\limsup_{n\to\infty}\mathbb{P}[\sup_{|t-s|\le\delta}|W_t^n - W_s^n| \ge \varepsilon] = 0 \tag{2.3}$$

for any $\varepsilon > 0$.

For (2.2), we have

$$(nt - [nt])\frac{X_{[nt]+1}}{\sqrt{n}\sigma} \xrightarrow{\mathbb{P}} 0$$

by Chebyshev's inequality, and

$$\frac{1}{\sqrt{n}\sigma}\sum_{i=1}^{[nt]} X_i \xrightarrow{d} \sqrt{t}N$$

by the Lindeberg-Lévy central limit theorem, where N is a random variable with the standard normal distribution, and so

$$W_t^n \xrightarrow{d} W_t$$

for any $t \in [0,1]$.

For $s < t$,

$$\begin{aligned}(W_s^n, W_t^n - W_s^n) &= \frac{1}{\sqrt{n}\sigma}(\sum_{i=1}^{[ns]} X_i, \sum_{[ns]+1}^{[nt]} X_i) + o_{\mathbb{P}}(1)\\ &\xrightarrow{d} (N_1, N_2)\end{aligned}$$

where N_1 and N_2 are independent normal random variables with mean 0 and variances s and $t - s$. Then using the continuous mapping theorem, we have

$$(W_s^n, W_t^n) \xrightarrow{d} (W_s, W_t).$$

Similarly, we can show (2.2).

Let $S_n = \sum_{i=1}^n X_i$. By the central limit theorem, for $k \le n$,

$$\mathbb{P}[|S_k| \ge \lambda\sigma\sqrt{n}] \le \mathbb{P}[|S_k| \ge \lambda\sigma\sqrt{k}] \to \mathbb{P}(|N| \ge \lambda) \le \frac{\mathbb{E}|N|^3}{\lambda^3}.$$

Set $E_i = \{\max_{1\le j<i}|S_j| < \lambda\sigma\sqrt{n} \le |S_i|\}$, $i = 1, 2, \cdots, n$, then $E_i \bigcap E_j = \emptyset$ for $i \ne j$, and

$$\bigcup_{i=1}^n E_i = \{\max_{1\le i\le n}|S_i| \ge \lambda\sigma\sqrt{n}\}.$$

We have

$$\begin{aligned}\mathbb{P}[\max_{1\le i\le n}|S_i| \ge \lambda\sigma\sqrt{n}] &\le \mathbb{P}[|S_n| \ge (\lambda - \sqrt{2})\sigma\sqrt{n}]\\ &\quad + \sum_{i=1}^{n-1}\mathbb{P}[E_i, |S_n| < (\lambda - \sqrt{2})\sigma\sqrt{n}],\end{aligned}$$

however for large n,

$$\begin{aligned}\sum_{i=1}^{n-1}\mathbb{P}[E_i, |S_n| < (\lambda-\sqrt{2})\sigma\sqrt{n}] &\le \sum_{i=1}^{n-1}\mathbb{P}[E_i, |S_n - S_i| \ge \sigma\sqrt{2n}]\\ &= \sum_{i=1}^{n-1}\mathbb{P}[E_i]\mathbb{P}[|S_n - S_i| \ge \sigma\sqrt{2n}]\\ &\le \frac{1}{2}\sum_{i=1}^{n-1}\mathbb{P}[E_i] \le \frac{1}{2}\mathbb{P}[\max_{1\le i\le n}|S_i| \ge \lambda\sigma\sqrt{n}],\end{aligned}$$

thus

$$\lim_{\lambda\to\infty}\limsup_{n\to\infty}\lambda^2\mathbb{P}[\max_{k\le n}|S_k| \ge \lambda\sigma\sqrt{n}] \le 2\lim_{\lambda\to\infty}\limsup_{n\to\infty}\lambda^2\mathbb{P}[|S_n| \ge (\lambda-\sqrt{2})\sigma\sqrt{n}] = 0. \tag{2.4}$$

Next step, we will prove that (2.4) implies (2.3).

Let $x \in \mathbb{C}[0,1]$, $0 = t_0 < t_1 < \cdots < t_v = 1$ with $\min_{1<i<v}(t_i - t_{i-1}) \ge \delta > 0$. Let $I_i = [t_{i-1}, t_i]$. For $s, t \in [0,1]$ with $t - \delta \le s \le t$, there are two cases.

If s and t lie in a same I_i, write

$$|x(s) - x(t)| \le |x(s) - x(t_{i-1})| + |x(t) - x(t_{i-1})|. \tag{2.5}$$

If s and t lie in adjacent I_i and I_{i+1}, write

$$|x(s) - x(t)| \le |x(s) - x(t_{i-1})| + |x(t_i) - x(t_{i-1})| + |x(t) - x(t_i)|. \tag{2.6}$$

From (2.5) and (2.6),

$$\sup_{|s-t|\le\delta}|x(s) - x(t)| \le 3\max_{1\le i\le v}\sup_{t_{i-1}\le s\le t_i}|x(s) - \dot{x}(t_{i-1})|. \tag{2.7}$$

For convenience, we take $t_i = im/n$ with integer m, $n\delta \le m \le n\delta + 1$, for $0 \le i \le v := [n/m] + 1$. Then

$$\begin{aligned}\mathbb{P}[\sup_{|s-t|\le\delta}|W_t^n - W_s^n| \ge 3\varepsilon] &\le \sum_{i=1}^{v}\mathbb{P}[\sup_{t_{i-1}\le s<t_i}|W_s^n - W_{t_{i-1}}^n| \ge \varepsilon]\\ &= \sum_{i=1}^{v}\mathbb{P}[\max_{(i-1)m\le k\le(i+1)m}\frac{|S_k - S_{(i-1)m}|}{\sigma\sqrt{n}} \ge \varepsilon]\\ &= v\mathbb{P}[\max_{k\le m}|S_k| \ge \varepsilon\sigma\sqrt{n}]\\ &\le \frac{2}{\delta}\mathbb{P}[\max_{k\le m}|S_k| \ge \frac{\varepsilon\sigma}{\sqrt{2\delta}}\sqrt{m}].\end{aligned}$$

Let $\lambda = \varepsilon/\sqrt{2\delta}$,

$$\mathbb{P}[\sup_{|s-t|\le\delta}|W_t^n - W_s^n| \ge 3\varepsilon] \le \frac{4\lambda^2}{\varepsilon^2}\mathbb{P}[\max_{k\le m}|S_k| \ge \lambda\sigma\sqrt{m}].$$

Then we can obtain (2.3) through (2.4). □

2.2.2 *Martingale invariance principle*

In the study of time series and statistics, dependent data is very common. To explore the large sample properties of dependent data, martingale approximation is a useful tool. Thus, martingale limit theorem may be necessary. Martingale central limit theory and invariance principle can be regarded as an extensions of the counterpart limit theory for sums of independent random variables.

For $n = 1, 2, \cdots$, let $\{X_{n,i}\}_{1\le i\le k_n}$ be a sequence of square-integrable martingale difference with respect to filter $\{\mathcal{F}_{n,i}\}_{1\le i\le k_n}$, the sub-$\sigma$−field $\mathcal{F}_{n,i}$ is generated by $X_{n,1}, \cdots, X_{n,i}$, and $S_{n,k} = \sum_{i=1}^{k} X_{n,i}$. Thus, $\{S_{n,i}, \mathcal{F}_{n,i}, 1 \le i \le k_n, n \ge 1\}$ is a martingale array. Lévy introduced the conditional variance

$$V_n^2 = \sum_{i=1}^{k_n} \mathbb{E}(X_{n,i}^2 | \mathcal{F}_{n,i})$$

as a counterpart of the variance in study of limit theory for sums of independent random variables. Moreover, from the Doob-Meyer decomposition of $U_n^2 = \sum_{i=1}^{k_n} X_{n,i}^2$, V_n^2 can be regarded as the compensator of U_n^2. It means that $\{U_{n,i}^2 - V_{n,i}^2\}$ is uniformly integrable martingale. From Theorem 2.23 in *Hall and Heyde (1980)*

$$\max_{1\le i\le k_n} |U_{n,i}^2 - V_{n,i}^2| \xrightarrow{\mathbb{P}} 0 \tag{2.8}$$

holds, where $U_{n,i}^2 = \sum_{j=1}^{i} X_{n,j}^2$, $V_{n,i} = \sum_{j=1}^{i} \mathbb{E}(X_{n,i}^2 | \mathcal{F}_{n,i})$. (2.8) means the conditional variance $V_{n,i}^2$ may be approximated by $U_{n,i}^2$. If $\{X_{n,i}\}_{1\le i\le k_n}$ is independent random variables with zero means, variances summing to 1, and for all $\varepsilon > 0$,

$$\max_{1\le i\le k_n} \mathbb{P}(|X_{n,i}| > \varepsilon) \to 0,$$

Raikov (1938) showed that

$$\sum_{i=1}^{k_n} X_{n,i} \xrightarrow{d} N(0,1)$$

iff

$$U_n^2 \xrightarrow{\mathbb{P}} 1. \tag{2.9}$$

However, when $\{X_{n,i}\}_{1\le i\le k_n}$ is a martingale difference sequence, (2.9) can not always hold due to (2.8). Usually, the limit of U_n^2 in probability is a random variable, it is quite different from the independent case. Thus, the central limit theorem for martingale is very interesting. *Chatterji (1974), Hall (1977), Rootzén (1977) and Aldous and Eagleson (1978)* studied the central limit theorems for martingales. Under some conditions, Aldous and Eagleson (1978) showed that

$$S_{n,k_n} \xrightarrow{d} N\eta, \tag{2.10}$$

where η is a random variable as the limit of U_n^2 in probability, and N is a standard normal variable independent of η. Obviously, $N\eta$ has characteristic function $\mathbb{E}\exp(-\frac{1}{2}\eta^2 t^2)$. (2.10) is equivalent to

$$\frac{S_{n,k_n}}{\sqrt{U_n^2}} \xrightarrow{d} N. \tag{2.11}$$

In this subsection, we will introduce the martingale invariance principle, or say martingale functional central limit theorem.

For each $t \in [0,1]$, similar to (2.1), we turn to study the weak convergence of random elements in $\mathbb{C}[0,1]$:

$$\zeta_n(t) = \frac{1}{\sqrt{U_n^2}}(S_{n,i} + \frac{(tU_n^2 - U_{n,i}^2)}{X_{n,i+1}^2} X_{n,i+1}), \quad \text{if } U_{n,i}^2 \le tU_n^2 < U_{n,i+1}^2.$$

However, $\zeta_n(t)$ is quite complex. We turn to study simpler case. Let

$$\beta_n(t) = \sum_{i=1}^{k_n} X_{n,i} 1_{\{U_{n,i}^2/U_n^2 \le t\}} \tag{2.12}$$

and

$$\xi_n(t) = \frac{1}{\sqrt{U_n^2}} \beta_n(t). \tag{2.13}$$

ξ_n and β_n are random elements in $\mathbb{D}[0,1]$.

Theorem 2.4. *(Martingale invariance principle)*

Suppose that

$$\lim_{n\to\infty} \mathbb{E}(\max_{1\le i\le k_n} X_{n,i}^2) = 0 \tag{2.14}$$

and there exist variables η_n adapted to the σ-fields $\mathcal{F}_{n,1}$ such that

$$U_n^2 - \eta_n \xrightarrow{\mathbb{P}} 0. \tag{2.15}$$

If

$$\lim_{\delta\to\infty} \liminf_{n\to\infty} \mathbb{P}(U_n^2 > \delta) = 0 \tag{2.16}$$

and

$$\sum_{i=1}^{k_n} |\mathbb{E}(X_{n,i}|\mathcal{F}_{n,i-1})| \xrightarrow{\mathbb{P}} 0, \tag{2.17}$$

then $\xi_n \Rightarrow W$, where W is a standard Brownian motion.

If

$$\frac{\max_{1\le i\le k_n} X_{n,i}^2}{U_n^2} \xrightarrow{\mathbb{P}} 0, \tag{2.18}$$

$$U_n^2 \xrightarrow{d} \eta \tag{2.19}$$

where η is a random variable and

$$\frac{1}{\sqrt{U_n^2}} \sum_{i=1}^{k_n} |\mathbb{E}(X_{n,i}|\mathcal{F}_{n,i-1})| \xrightarrow{\mathbb{P}} 0, \tag{2.20}$$

then

$$(\beta_n, U_n^2) \Rightarrow ((\eta')^{1/2} W, \eta'), \tag{2.21}$$

where η' is a copy of η, independent of W.

Proof. We first prove that (2.14), (2.15) and (2.18)–(2.20) are sufficient for (2.21). Note that (2.21) is equivalent to the pair of conditions:

(i) for all sequences $0 = t_0 < t_1 < \cdots < t_p \leq 1$ and real numbers $z_1, z_2, \cdots, z_p$, s,

$$\lim_{n\to\infty} |\mathbb{E}\exp[\mathrm{i}\sum_{k=1}^{p} z_k(\beta_n(t_k)-\beta_n(t_{k-1}))+\mathrm{i}sU_n^2]-\mathbb{E}\exp[-\frac{1}{2}\eta\sum_{k=1}^{p} z_k^2(t_k-t_{k-1})+\mathrm{i}s\eta]| = 0; \tag{2.22}$$

(ii) the sequence of random elements $\{\beta_n, U_n^2, n \geq 1\}$ is tight.

Suppose first that for some $\lambda > 0$, $\mathbb{P}(\eta > \lambda) = 0$. η_n can be chosen such that $\mathbb{P}(\eta_n > 2\lambda) = 0$ and (2.15) is satisfied. Fix $\Delta > 0, \delta > 0$, define

$$\eta_n(\delta) = \delta 1_{\{\eta_n \leq \delta\}} + \eta_n 1_{\{\eta_n > \delta\}}$$

and

$$\beta_n(\delta, t) = \sum_{i=1}^{k_n} X_{n,i} 1_{\{U_{n,i-1}^2/\eta_n(\delta) \leq t\}}.$$

Define sets

$$F_n = \{\max_{i \leq k_n} X_{n,i}^2 \leq \frac{\Delta}{2} U_n^2;\ |U_n^2/\eta_n(\delta) - 1| \leq \frac{\Delta}{2}\}$$

and

$$G_n = \{|U_n^2 - \eta_n| \leq \Delta;\ \eta_n \leq \delta\}.$$

We need the following lemma.

Lemma 2.1. *For all $\varepsilon > 0$ and $t \in [0, 1]$,*

$$\lim_{\delta\to 0} \limsup_{n\to\infty} \mathbb{P}(|\beta_n(t) - \beta_n(\delta, t)| > \varepsilon) = 0. \tag{2.23}$$

Proof of Lemma 2.1. Note that

$$|U_{n,i}^2/\eta_n - U_{n,i-1}^2/\eta_n(\delta)| \leq X_{n,i}^2/\eta_n + |\eta_n/\eta_n(\delta) - 1|,$$

and so,

$$1_{\{U_{n,i-1}^2/\eta_n(\delta) \leq t-\Delta\}} \leq 1_{\{U_{n,i-1}^2/U_n^2 \leq t\}} \leq 1_{\{U_{n,i-1}^2/\eta_n(\delta) \leq t+\Delta\}}$$

on the set F_n. Hence

$$\begin{aligned} &|\beta_n(t) - \beta_n(\delta, t)| \\ &= |\sum_{i=1}^{k_n} X_{n,i}\{1_{\{U_{n,i-1}^2/U_n^2 \leq t\}} - 1_{\{U_{n,i-1}^2/\eta_n(\delta) \leq t\}}\}| \\ &\leq \max_{1\leq m<k\leq k_n} |\sum_{i=m}^{k} X_{n,i} 1_{\{t-\Delta \leq U_{n,i-1}^2/\eta_n(\delta) \leq t+\Delta\}}| \\ &\leq 2 \max_{1\leq m\leq k_n} |\sum_{i=1}^{m} X_{n,i} 1_{\{t-\Delta \leq U_{n,i-1}^2/\eta_n(\delta) \leq t+\Delta\}}| \end{aligned}$$

on F_n. Furthermore, on the set G_n,

$$|\beta_n(t)| \vee |\beta_n(\delta,t)| \le \max_{1\le m\le k_n} |\sum_{i=1}^m X_{n,i}| = \max_{1\le m\le k_n} |\sum_{i=1}^m X_{n,i} 1_{\{U^2_{n,i-1}\le \Delta+\delta\}}|.$$

Therefore,

$$\begin{aligned}
&\mathbb{P}(|\beta_n(t) - \beta_n(\delta,t)| > \varepsilon) \\
&\le \mathbb{P}(\max_{1\le m\le k_n} |\sum_{i=1}^m X_{n,i} 1_{\{U^2_{n,i-1}\le \Delta+\delta\}}| \ge \frac{\varepsilon}{2}) \\
&+\mathbb{P}(\max_{1\le m\le k_n} |\sum_{i=1}^m X_{n,i} 1_{\{t-\Delta\le \{U^2_{n,i-1}/\eta_n(\delta)\le t+\Delta\}}| \ge \frac{\varepsilon}{4}) + \mathbb{P}(F_n^c \bigcap G_n^c).
\end{aligned}$$

(2.15) and (2.18) imply that the first two terms on the right hand side are $o(1)$. The last term,

$$\begin{aligned}
&\mathbb{P}(F_n^c \bigcap G_n^c) \\
&\le \mathbb{P}(\max_{i\le k_n} X^2_{n,i} > \frac{\Delta}{2} U_n^2) + \mathbb{P}(|U_n^2/\eta_n(\delta) - 1| > \frac{\Delta}{2};\ \eta_n > \delta),
\end{aligned}$$

where

$$\mathbb{P}(|U_n^2/\eta_n(\delta) - 1| > \frac{\Delta}{2};\ \eta_n > \delta) \le \mathbb{P}(|U_n^2 - \eta_n| > \frac{\Delta\delta}{2}) \to 0$$

by (2.15). Hence,

$$\mathbb{P}(F_n^c \bigcap G_n^c) = o(1). \tag{2.24}$$

Set

$$P_{n,m} = \sum_{i=1}^m X_{n,i} 1_{\{t-\Delta\le U^2_{n,i-1}/\eta_n(\delta)\le t+\Delta\}},$$

$$Q_{n,m} = \sum_{i=1}^m X_{n,i} 1_{\{U^2_{n,i-1}\le \Delta+\delta\}}.$$

Obviously, $(P_{n,m}, \mathcal{F}_{n,m})_{m\ge 1}$, $(Q_{n,m}, \mathcal{F}_{n,m})_{m\ge 1}$ are martingales. Applying Kolmogorov's inequality for martingales, we have

$$\begin{aligned}
&\mathbb{P}(\max_{1\le m\le k_n} |\sum_{i=1}^m X_{n,i} 1_{\{t-\Delta\le U^2_{n,i-1}/\eta_n(\delta)\le t+\Delta\}} \ge \frac{\varepsilon}{4}) \\
&\le 16\varepsilon^{-2} \mathbb{E}[\sum_{i=1}^{k_n} X^2_{n,i} 1_{\{t-\Delta\le U^2_{n,i-1}/\eta_n(\delta)\le t+\Delta\}}] \\
&\le 32\varepsilon^{-2} \frac{\lambda}{\eta_n(\delta)} \mathbb{E}[\sum_{i=1}^{k_n} X^2_{n,i} 1_{\{t-\Delta\le U^2_{n,i-1}/\eta_n(\delta)\le t+\Delta\}}]
\end{aligned}$$

$$\leq 32\varepsilon^{-2}\lambda\mathbb{E}[2\Delta + \frac{\max_{i\leq k_n} X_{n,i}^2}{\delta}]$$
$$= 64\varepsilon^{-2}\lambda\Delta + o(1).$$

Similarly, we can obtain

$$\mathbb{P}(\max_{1\leq m\leq k_n} |\sum_{i=1}^{m} X_{n,i}1_{\{U_{n,i-1}^2\leq\Delta+\delta\}} \geq \frac{\varepsilon}{2}) \leq 4\varepsilon^{-2}(\Delta+\delta) + o(1).$$

Hence

$$\mathbb{P}(|\beta_n(t) - \beta_n(\delta, t)| > \varepsilon) \leq 64\varepsilon^{-2}\lambda\Delta + 4\varepsilon^{-2}(\Delta+\delta) + o(1).$$

Let $n \to \infty$ and then $\Delta \to 0$, $\delta \to 0$ to establish (2.23). □

In view of (2.15) and (2.23), (2.22) will hold if we show that

$$\lim_{n\to\infty} |\mathbb{E}\exp[\mathrm{i}\sum_{k=1}^{p} z_k(\beta_n(\delta, t_k) - \beta_n(\delta, t_{k-1})) + \mathrm{i}s\eta_n]$$
$$-\mathbb{E}\exp[-\frac{1}{2}\eta\sum_{k=1}^{p} z_k^2(t_k - t_{k-1}) + \mathrm{i}s\eta]| = 0. \tag{2.25}$$

Define

$$A_{n,i} = A_{n,i}(\delta) = X_{n,i}\sum_{k=1}^{p} z_k 1_{\{t_{k-1}<U_n^2/\eta_n(\delta)\leq t_k\}}, \quad 1\leq i\leq k_n,$$

$$B_n^2 = B_n^2(\delta) = \sum_{i=1}^{k_n} A_{n,i}^2 = \sum_{i=1}^{k_n} X_{n,i}^2 \sum_{k=1}^{p} z_k^2 1_{\{t_{k-1}<U_n^2/\eta_n(\delta)\leq t_k\}}$$

and

$$T_n = T_n(\delta) = \prod_{i=1}^{k_n}(1 + \mathrm{i}A_{n,i}),$$

$$W_n = W_n(\delta) = \exp(-\frac{1}{2}B_n^2 + \sum_{i=1}^{k_n} r(A_{n,i}) + \mathrm{i}s\eta_n).$$

Write $e^{\mathrm{i}x} = (1+x)\exp(-\frac{1}{2}x^2 + r(x))$, where $|r(x)| \leq |x|^3$ for $|x| < 1$. Let

$$w_n = \exp(-\frac{1}{2}\sigma^2\eta_n + \mathrm{i}s\eta_n),$$

$$\sigma^2 = \sum_{k=1}^{p} z_k^2(t_k - t_{k-1})$$

and $\psi = \mathbb{E}\exp(-\frac{\eta}{2}\sigma^2 + \mathrm{i}s\eta)$. Then

$$\mathbb{E}\exp[\mathrm{i}\sum_{k=1}^{p} z_k(\beta_n(\delta, t_k) - \beta_n(\delta, t_{k-1})) + \mathrm{i}s\eta_n]$$

$$-\mathbb{E}\exp[-\frac{1}{2}\eta\sum_{k=1}^{p}z_k^2(t_k-t_{k-1})+is\eta]$$
$$=\mathbb{E}(T_nW_n)-\psi=\mathbb{E}T_n(W_n-w_n)+\mathbb{E}(T_n-1)w_n+\mathbb{E}(w_n)-\psi$$

and

$$|\mathbb{E}(T_n-1)w_n|\le\mathbb{E}|\mathbb{E}(T_n|\mathcal{F}_{n,1})-1|=0.$$

Moreover, $\mathbb{E}(w_n)\to\psi$ as $n\to\infty$, since (2.15), (2.18) and the functions $f(u)=\exp(-\frac{1}{2}\sigma^2u)$ and $g(u)=\exp(isu)$ are uniformly bounded and continuous on $[0,\infty)$. Now, we show

$$\lim_{\delta\to0}\limsup_{n\to\infty}\mathbb{E}|T_n(\delta)(W_n(\delta)-w_n)|=0. \tag{2.26}$$

First let us prove

Lemma 2.2. *For all $\varepsilon>0$,*

$$\lim_{\delta\to0}\limsup_{n\to\infty}\mathbb{P}(|W_n(\delta)-w_n|>\varepsilon)=0. \tag{2.27}$$

Proof of Lemma 2.2. Let $z=\max_{1\le k\le p}|z_k|$. We have

$$|\sum_{i=1}^{k_n}r(A_{n,i})|\le\sum_{i=1}^{k_n}|A_{n,i}|^3\le B_n^2\max_{i\le k_n}|A_{n,i}|$$
$$\le z^3U_n^2\max_{i\le k_n}|X_{n,i}|\xrightarrow{\mathbb{P}}0$$

as $n\to\infty$. Furthermore,

$$\sum_{k=1}^{p}z_k^2\sum_{i=1}^{k_n}X_{n,i}^2 1_{\{t_{k-1}+\Delta<U_{n,i}^2/U_n^2\le t_k-\Delta\}}$$
$$\le\sum_{k=1}^{p}z_k^2\sum_{i=1}^{k_n}X_{n,i}^2 1_{\{t_{k-1}-\Delta<U_{n,i}^2/U_n^2\le t_k+\Delta\}}.$$

In view of (2.18), for each $t\in[0,1]$,

$$D_n^2(t):=\frac{1}{U_n^2}\sum_{i=1}^{k_n}X_{n,i}^2 1_{\{U_{n,i}^2/U_n^2\le t\}}\xrightarrow{\mathbb{P}}t.$$

If $t>1$, then $D_n^2(t)=1$ and if $t<0$, $D_n^2(t)=0$. If Δ is so small that each $t_{k-1}+\Delta\le t_k-\Delta$, then we have

$$\sum_{k=1}^{p}z_k^2[D_n^2(t_k-\Delta)-D_n^2(t_{k-1}+\Delta)]\le B_n^2/U_n^2\le\sum_{k=1}^{p}z_k^2[D_n^2(t_k+\Delta)-D_n^2(t_{k-1}-\Delta)].$$

Hence on the set $F_n\bigcap\{\max_{0\le k\le p}|\sum_{k=1}^{p}[D_n^2(t_k\pm\Delta)-(t_k\pm\Delta)]|\le\Delta\}\bigcap\{U_n^2\le2\lambda\}$,

$$U_n^2\sigma^2-8pz^2\Delta\lambda\le B_n^2\le U_n^2\sigma^2+8pz^2\Delta\lambda.$$

Let $\varepsilon > 0$ and choose Δ so small that $8pz^2\Delta\lambda < \varepsilon$. Then

$$\begin{aligned}
&\mathbb{P}(|B_n^2 - U_n^2\sigma^2| > \varepsilon; F_n) \\
&\le \mathbb{P}(\max_{0\le k\le p} |D_n^2(t_k \pm \Delta) - (t_k \pm \Delta)| > \Delta) + P(U_n^2 > 2\lambda),
\end{aligned}$$

where $\mathbb{P}(U_n^2 > 2\lambda) \to 0$, since $U_n^2 \xrightarrow{d} \eta \le \lambda$ *a.s.*, and so for all sufficiently small Δ and all $\delta > 0$,

$$\lim_{n\to\infty} \mathbb{P}(|B_n^2(\delta) - U_n^2\sigma^2| > \varepsilon;\ F_n(\delta,\Delta)) = 0.$$

On the set G_n, $U_n^2 \le \Delta + \delta$ and so

$$|B_n^2 - U_n^2\sigma^2| \le 2z^2U_n^2 \le 2z^2(\Delta + \delta).$$

Choose Δ and δ so small that $2z^2(\Delta+\delta) < \varepsilon$. Then for all n,

$$\mathbb{P}(|B_n^2 - U_n^2\sigma^2| > \varepsilon;\ G_n(\delta,\Delta)) = 0.$$

Thus, for all sufficiently small Δ and δ,

$$\mathbb{P}(|B_n^2 - U_n^2\sigma^2| > \varepsilon) = o(1).$$

Then we have (2.27) by (2.15). □

Now we show (2.26). Let

$$C_n = \sum_{i=1}^{k_n} A_{n,i} = \sum_{k=1}^{p} z_k(\beta_n(\delta,t_k) - \beta_n(\delta,t_{k-1})),$$

$$I_n = I_n(\delta) = \exp(\mathrm{i}C_n + \mathrm{i}s\eta_n).$$

For $\varepsilon > 0$,

$$\begin{aligned}
&\mathbb{E}|T_n(\delta)(W_n(\delta) - w_n)| \\
&\le (\mathbb{E}|T_n|^2)^{1/2}\varepsilon + \int_{\{|W_n - w_n|>\varepsilon\}} |T_n(W_n - w_n)| d\mathbb{P} \\
&\le (\mathbb{E}|T_n|^2)^{1/2}\varepsilon + \int_{\{|W_n - w_n|>\varepsilon\}} (|I_n| + |T_n w_n|) d\mathbb{P} \\
&\le (\mathbb{E}|T_n|^2)^{1/2}\varepsilon + \mathbb{P}(|W_n - w_n| > \varepsilon) + (\mathbb{E}(T_n^2))^{1/2}(\mathbb{P}(|W_n - w_n| > \varepsilon))^{1/2} \\
&= (\mathbb{E}|T_n|^2)^{1/2}(\varepsilon + (\mathbb{P}(|W_n - w_n| > \varepsilon))^{1/2}) + \mathbb{P}(|W_n - w_n| > \varepsilon).
\end{aligned}$$

Let $J_n = \max\{i \le k_n | U_{n,i-1}^2/\eta_n(\delta) \le 1\}$. Since $A_{n,i}^2 \le t^2X_{n,i}^2$,

$$\begin{aligned}
\mathbb{E}|T_n|^2 = \mathbb{E}\prod_{i=1}^{k_n}(1 + A_{n,i}^2) &\le \mathbb{E}\exp(z^2U_{n,J_n-1}^2)(1 + z^2X_{n,J_n}^2) \\
&\le \mathbb{E}\exp(z^2\eta_n(\delta))(1 + z^2\max_{i\le k_n} X_{n,i}^2) \\
&\le \exp(2\lambda z^2)(1 + z^2\mathbb{E}(\max_{i\le k_n} X_{n,i}^2)) \\
&\to \exp(2\lambda z^2) \quad \text{as } n\to\infty.
\end{aligned}$$

Combining Lemma 2.2, we get (2.26). Hence, we obtain (2.22) under the conditions (2.14), (2.15), (2.18) and (2.19) for some $\lambda > 0$ with $\mathbb{P}(\eta > \lambda) = 0$.

Now let us consider the case of an arbitrary η. Let $\lambda > 0$ be a continuity point of distribution function of η, and define $\widetilde{X}_{n,i} = X_{n,i}1_{\{U^2_{n,i-1} \le \lambda\}}$, $\widetilde{S}_{n,j} = \sum_{i=1}^{j} \widetilde{X}_{n,i}$, $\widetilde{U}_n^2 = \sum_{i=1}^{k_n} \widetilde{X}^2_{n,i}$, $\widetilde{\eta}_n = \eta_n 1_{\{\eta_n \le \lambda\}} + \lambda 1_{\{\eta_n > \lambda\}}$, $\widetilde{\eta} = \eta 1_{\{\eta \le \lambda\}} + \lambda 1_{\{\eta > \lambda\}}$. Obviously, $\widetilde{X}^2_{n,i} \le X^2_{n,i}$ implies

$$\lim_{n\to\infty} \mathbb{E}(\max_{1\le i\le k_n} \widetilde{X}^2_{n,i}) = 0. \tag{2.28}$$

Moreover, $|\widetilde{U}_n^2 - \widetilde{\eta}_n| \le |U_n^2 - \eta_n| + \max_{i\le k_n} X^2_{n,i}$ and so

$$\widetilde{U}_n^2 - \widetilde{\eta}_n \xrightarrow{\mathbb{P}} 0. \tag{2.29}$$

Note that

$$\frac{\max_{i\le k_n} \widetilde{X}^2_{n,i}}{\widetilde{U}_n^2} \le \max(\frac{\max_{i\le k_n} X^2_{n,i}}{U_n^2}, \frac{\max_{i\le k_n} X^2_{n,i}}{\lambda}) \xrightarrow{\mathbb{P}} 0, \tag{2.30}$$

since λ is a continuity point of the distribution function of η,

$$\widehat{U}_n^2 := U_n^2 1_{\{U_n^2 \le \lambda\}} + \lambda 1_{\{U_n^2 > \lambda\}} \xrightarrow{d} \widetilde{\eta}.$$

Furthermore,

$$\widehat{U}_n^2 \le \widetilde{U}_n^2 \le \widehat{U}_n^2 + \max_{i\le k_n} X^2_{n,i},$$

we have

$$\widetilde{U}_n^2 \xrightarrow{d} \widetilde{\eta}. \tag{2.31}$$

Define $\widetilde{\beta}_n$ for the martingale $\{(\widetilde{S}_{n,i}, \mathcal{F}_{n,i}), 1 \le i \le k_n\}$ in the same way that we defined β_n for $\{(S_{n,i}, \mathcal{F}_{n,i}), 1 \le i \le k_n\}$. From (2.28), (2.29), (2.30) and (2.31), the proof given above implies that

$$\lim_{n\to\infty} |\mathbb{E}\exp[\mathrm{i}\sum_{k=1}^{p} z_k(\widetilde{\beta}_n(t_k) - \widetilde{\beta}_n(t_{k-1})) + \mathrm{i}s\widetilde{U}_n^2] - \mathbb{E}\exp[-\frac{1}{2}\eta\sum_{k=1}^{p} z_k^2(t_k - t_{k-1}) + \mathrm{i}s\widetilde{\eta}]| = 0. \tag{2.32}$$

Since

$$\limsup_{n\to\infty} \mathbb{P}(U_n^2 > \lambda) \to 0$$

as $\lambda \to \infty$, (2.32) implies (2.22) under the conditions (2.14), (2.15), (2.18) and (2.19).

Now we will prove the tightness of $\{\beta_n, n \ge 1\}$ under the conditions (2.14), (2.15), (2.18) and (2.19). Suppose first that for some $\lambda > 0$, $\mathbb{P}(\eta > \lambda) = 0$. η_n can be chosen such that $\mathbb{P}(\eta_n > 2\lambda) = 0$. We will obtain the tightness of $\{\beta_n, n \ge 1\}$ through showing

$$\lim_{h\to 0} \limsup_{n\to\infty} \sum_{kh<1} \mathbb{P}(\sup_{kh<t<(k+1)h} |\beta_n(t) - \beta_n(kh)| > \varepsilon) = 0 \tag{2.33}$$

for all $\varepsilon > 0$. On the set F_n,

$$\max_{i \le k_n} |U_{n,i}^2/U_n^2 - U_{n,i-1}^2/\eta_n(\delta)| \le \Delta$$

and so

$$\sup_{kh<t<(k+1)h} |\beta_n(t) - \beta_n(kh)| \le 2 \max_{m \le k_n} |\sum_{i=1}^m X_{n,i} 1_{\{kh-\Delta<U_{n,i-1}^2/\eta_n(\delta)\le(k+1)h+\Delta\}}|$$

on F_n. Moreover,

$$\sup_{kh<t<(k+1)h} |\beta_n(t) - \beta_n(kh)| \le 2 \max_{m \le k_n} |\sum_{i=1}^m X_{n,i} 1_{\{U_{n,i-1}^2 \le \Delta+\delta\}}|$$

on the set G_n.

Therefore

$$\mathbb{P}(\sup_{kh<t<(k+1)h} |\beta_n(t) - \beta_n(kh)| > \varepsilon) \tag{2.34}$$

$$\le \mathbb{P}(\max_{m \le k_n} |\sum_{i=1}^m X_{n,i} 1_{\{kh-\Delta<U_{n,i-1}^2/\eta_n(\delta)\le(k+1)h+\Delta\}}| > \varepsilon/2) \tag{2.35}$$

$$+\mathbb{P}(\max_{m \le k_n} |\sum_{i=1}^m X_{n,i} 1_{\{U_{n,i-1}^2 \le \Delta+\delta\}}| > \varepsilon/2) + \mathbb{P}(F_n^c \bigcap G_n^c). \tag{2.36}$$

Consider the term in (2.35). Set $0 < \varepsilon_1 < \varepsilon/8$ and $C > \varepsilon$,

$$M_{n,m} = \sum_{i=1}^m X_{n,i} 1_{\{kh-\Delta<U_{n,i-1}^2/\eta_n(\delta)\le(k+1)h+\Delta\}}, \quad 1 \le m \le k_n,$$

$$L_n = \{|\beta_n(kh-\Delta) - \beta_n(\delta, kh-\Delta)| < \varepsilon_1; |\beta_n((k+1)h+\Delta) - \beta_n(\delta, (k+1)h+\Delta)| < \varepsilon_1\}$$

and $M_n = M_{n,k_n}$. Obviously, $\{(M_{n,i}, \mathcal{F}_{n,i}), 1 \le i \le k_n\}$ is a martingale, and by Kolmogorov's inequality,

$$\begin{aligned}
&\mathbb{P}(\max_{m \le k_n} |\sum_{i=1}^m X_{n,i} 1_{\{kh-\Delta<U_{n,i-1}^2/\eta_n(\delta)\le(k+1)h+\Delta\}}| > \varepsilon/2) \\
&\le \frac{4}{\varepsilon} \int_{\{|M_n|>\varepsilon/4\}} |M_n| d\mathbb{P} \\
&\le \frac{4}{\varepsilon} \int_{\{|M_n|>C\}} |M_n| d\mathbb{P} + \frac{4}{\varepsilon} \int_{L_n^c} |M_n| d\mathbb{P} + \frac{4}{\varepsilon} \int_{L_n \bigcap \{C \ge |M_n| \ge \varepsilon/4\}} |M_n| d\mathbb{P}.
\end{aligned}$$

Furthermore,

$$\begin{aligned}
&\frac{4}{\varepsilon} \int_{\{|M_n|>C\}} |M_n| d\mathbb{P} \\
&\le \frac{4}{\varepsilon C} \mathbb{E} M_n^2 \\
&= \frac{4}{\varepsilon C} \mathbb{E}[\sum_{i=1}^{k_n} X_{n,i}^2 1_{\{kh-\Delta<U_{n,i-1}^2/\eta_n(\delta)\le(k+1)h+\Delta\}}]
\end{aligned}$$

$$\leq \frac{8\lambda}{\varepsilon C}\mathbb{E}[\frac{1}{\eta_n(\delta)}\sum_{i=1}^{k_n} X_{n,i}^2 1_{\{kh-\Delta<U_{n,i-1}^2/\eta_n(\delta)\leq(k+1)h+\Delta\}}]$$
$$\leq \frac{8\lambda}{\varepsilon C}\mathbb{E}[C' + \frac{\max_{i\leq k_n} X_{n,i}^2}{\delta}] = \frac{8C'\lambda}{\varepsilon C} + o(1)$$

where $C' = (k+1)h + \Delta$, and

$$\frac{4}{\varepsilon}\int_{L_n^c} |M_n| d\mathbb{P} \leq \frac{4}{\varepsilon}(\mathbb{P}(L_n^c)EM_n^2)^{1/2} + o(1).$$

On L_n

$$\begin{aligned}|M_n| &= |\beta_n(\delta,(k+1)h+\Delta) - \beta_n(\delta, kh-\Delta)| \\ &\leq |\beta_n((k+1)h+\Delta) - \beta_n(kh-\Delta)| + 2\varepsilon_1 \\ &=: |M_n'| + 2\varepsilon_1,\end{aligned}$$

then

$$\frac{4}{\varepsilon}\int_{L_n\bigcap\{C\geq|M_n|\geq\varepsilon/4\}} |M_n| d\mathbb{P} \leq \frac{4}{\varepsilon}(2\varepsilon_1 + \int_{\{C+2\varepsilon_1\geq|M_n'|\geq\varepsilon/4-2\varepsilon_1\}} |M_n'| d\mathbb{P}).$$

The convergence of finite dimensional distributions of β_n implies that

$$\lim_{n\to\infty}\int_{\{C+2\varepsilon\geq|M_n'|\geq\varepsilon/4-2\varepsilon\}} |M_n'| d\mathbb{P} = \mathbb{E}[\int_{\varepsilon/4-2\varepsilon_1}^{C+2\varepsilon_1} x d\mathbb{P}(|N(0,\eta(h+2\Delta)| \leq x)]. \quad (2.37)$$

Now consider the first term of (2.36). By Kolmogorov's inequality,

$$\mathbb{P}(\max_{m\leq k_n} |\sum_{i=1}^{m} X_{n,i} 1_{\{U_{n,i-1}^2\leq\Delta+\delta\}}| > \frac{\varepsilon}{2}) \leq \frac{4}{\varepsilon^2}(\Delta+\delta) + o(1).$$

Combining (2.35), (2.37) with (2.24), we see that

$$\begin{aligned}&\limsup_{n\to\infty}\mathbb{P}(\sup_{kh<t<(k+1)h} |\beta_n(t) - \beta_n(kh)| > \varepsilon) \\ &\leq \frac{16\lambda}{\varepsilon C} + \frac{4}{\varepsilon}(\limsup_{n\to\infty}\mathbb{P}(L_n^c)4\lambda)^{1/2} + \frac{\varepsilon^2}{4}(\Delta+\delta) \\ &\quad + \frac{\varepsilon}{4}\mathbb{E}[\int_{\varepsilon/4-2\varepsilon_1}^{C+2\varepsilon_1} x d\mathbb{P}(|N(0,\eta(h+2\Delta)| \leq x)].\end{aligned}$$

Let $\delta \to 0$ and then $\varepsilon_1 \to 0$, $\Delta \to 0$ and $C \to \infty$, we obtain

$$\begin{aligned}&\limsup_{n\to\infty}\mathbb{P}(\sup_{kh<t<(k+1)h} |\beta_n(t) - \beta_n(kh)| > \varepsilon) \\ &\leq \frac{\varepsilon}{4}\mathbb{E}[\int_{\varepsilon/4}^{\infty} x d\mathbb{P}(|N(0,\eta h| \leq x)] = \frac{\varepsilon}{4}\mathbb{E}[(\frac{2}{\pi})^{1/2}(\eta h)^{1/2}\exp(-\frac{1}{2}(\frac{\varepsilon}{4})^2(\eta h)^{-1})].\end{aligned}$$

Hence

$$\limsup_{n\to\infty}\sum_{kh\leq 1}\mathbb{P}(\sup_{kh<t<(k+1)h} |\beta_n(t) - \beta_n(kh)| > \varepsilon)$$

$$\leq \frac{\varepsilon}{4}\mathbb{E}[(\frac{2}{\pi})^{1/2}(\eta h)^{1/2}\exp(-\frac{1}{2}(\frac{\varepsilon}{4})^2(\eta h)^{-1})] + \delta_h,$$

where $\delta_h \to 0$ as $h \to 0$.

The integrand of the expectation on the right converges a.s. to 0 as $h \to 0$ and is dominated by $(4/\varepsilon)T \leq (4/\varepsilon)\lambda$. Hence the expectation itself converges to 0, and this establishes (2.33). This proves the tightness of β_n in the case where η is essentially bounded. A proof in the more general case follows via a truncation argument like that used in the proof of (2.22). Hence, we obtain (2.21) under the conditions (2.14), (2.15), (2.18) and (2.19).

Now we will prove the theorem. Let $Y_{n,i} = X_{n,i} - \mathbb{E}(X_{n,i}|\mathcal{F}_{n,i-1})$, $T_{n,i} = \sum_{k=1}^{i} Y_{n,k}$, $V_{n,i}^2 = \sum_{k=1}^{i} Y_{n,k}^2$ and $V_n^2 = V_{n,k_n}^2$. Define

$$\eta_n(t) = \sum_{i=1}^{[nt]} Y_{n,i}1_{\{V_{n,i}^2/V_n^2 \leq t\}}.$$

(2.14) implies

$$\lim_{n\to\infty} \mathbb{E}(\max_{1\leq i\leq k_n} Y_{n,i}^2) = 0 \tag{2.38}$$

and so the martingales $(\{T_{n,i}, \mathcal{F}_{n,i}), 1 \leq i \leq k_n\}$, $n \geq 1$, satisfy the analogue of (2.14).

$$\frac{\max_{i\leq k_n}|U_{n,i}^2 - V_{n,i}^2|}{U_n^2} \leq 2\frac{(\sum_{i=1}^{k_n} X_{n,i}^2)^{1/2}(\sum_{i=1}^{k_n}|\mathbb{E}(X_{n,i}|\mathcal{F}_{n,i-1})|^2)^{1/2}}{U_n^2}$$
$$+\frac{(\sum_{i=1}^{k_n}|\mathbb{E}(X_{n,i}|\mathcal{F}_{n,i-1})|)^2}{U_n^2}$$
$$= 2\theta_n + \theta_n^2$$

where $\theta_n = \sqrt{\frac{(\sum_{i=1}^{k_n}|\mathbb{E}(X_{n,i}|\mathcal{F}_{n,i-1})|)^2}{U_n^2}}$. By (2.20),

$$\theta_n \xrightarrow{\mathbb{P}} 0$$

and so

$$\frac{\max_{i\leq k_n}|U_{n,i}^2 - V_{n,i}^2|}{U_n^2} \xrightarrow{\mathbb{P}} 0 \tag{2.39}$$

and furthermore,

$$|U_n^2 - V_n^2| \xrightarrow{\mathbb{P}} 0.$$

Hence, the analogues of (2.15) and (2.19) hold for the martingale $(\{T_{n,i}, \mathcal{F}_{n,i}), 1 \leq i \leq k_n\}$. From the above proof, we obtain (2.21) is equivalent to the pair of the conditions:

(iii) for all sequence $0 = t_0 < t_1 < \cdots < t_p \leq 1$ and real numbers $z_1, z_2, \cdots, z_p$, s,

$$\lim_{n\to\infty}|\mathbb{E}\exp[\mathrm{i}\sum_{k=1}^{p} z_k(\eta_n(t_k)-\eta_n(t_{k-1}))+\mathrm{i}sV_n^2]-\mathbb{E}\exp[-\frac{1}{2}\eta\sum_{k=1}^{p} z_k^2(t_k-t_{k-1})+\mathrm{i}s\eta]| = 0;$$

(iv) the sequence of random elements $\{\eta_n, n \geq 1\}$ is tight.

The theorem will be obtained by

$$\sup_{t\in[0,1]} |\eta_n(t) - \beta_n(t)| \xrightarrow{\mathbb{P}} 0. \tag{2.40}$$

On the set $\{\sup_{i\leq k_n} |U_{n,i}^2/U_n^2 - V_{n,i}^2/V_n^2| \leq \delta\}$ we have

$$\sup_{t\in[0,1]} |\eta_n(t) - \beta_n(t)| \leq \sup_{s,w\in[0,1];|s-w|\leq\delta} |\beta_n(w) - \beta_n(s)| + \frac{\sum_{i=1}^{k_n} |\mathbb{E}(X_{n,i}|\mathcal{F}_{n,i-1})|}{\sqrt{U_n^2}}$$

and for any $\varepsilon > 0$,

$$\begin{aligned}\mathbb{P}(\sup_{t\in[0,1]} |\eta_n(t) - \beta_n(t)| > \varepsilon) \leq \mathbb{P}(&\sup_{s,w\in[0,1];|s-w|\leq\delta} |\beta_n(w) - \beta_n(s)| > \varepsilon/2)\\ &+\mathbb{P}(\frac{\sum_{i=1}^{k_n} |\mathbb{E}(X_{n,i}|\mathcal{F}_{n,i-1})|}{\sqrt{U_n^2}} > \varepsilon/2)\\ &+\mathbb{P}(\sup_{i\leq k_n} |U_{n,i}^2/U_n^2 - V_{n,i}^2/V_n^2| > \delta).\end{aligned}$$

By (2.20), (2.39) and the tightness of β_n, the theorem is proved. □

Corollary 2.1. *Suppose that (2.14) holds,*

$$U_n^2 \xrightarrow{\mathbb{P}} \eta, \tag{2.41}$$

and either

$$\eta \textit{ is measurable in } \bigcap_{n=1}^{\infty} \mathcal{F}_{n,1} \tag{2.42}$$

or

$$k_n \uparrow \infty \textit{ and } \mathcal{F}_{n,i} \subseteq \mathcal{F}_{n+1,i} \textit{ for all } i \leq k_n. \tag{2.43}$$

Then (2.21) holds.

Proof. We need to prove that (2.14), (2.41) and (2.42) or (2.43) are sufficient for (2.21). Taking $\eta_n = \eta$, (2.20) implies (2.18), so (2.21) holds. If (2.43) is true, choose integers $l_n \leq k_n$ such that $l_n \uparrow \infty$, and

$$l_n \frac{\max_{i\leq k_n} |X_{n,i}|}{\sqrt{U_n^2}} \xrightarrow{\mathbb{P}} 0.$$

η is measurable in the $\sigma-$field generated by $\bigcup_{n=1}^{\infty} \mathcal{F}_{n,k_n} = \bigcup_{n=1}^{\infty} \mathcal{F}_{n,l_n}$. For each $\varepsilon > 0$, we can find an n and an $\mathcal{F}_{n,l_n}$-measurable variable ζ such that

$$\mathbb{P}(|\eta - \zeta| > \varepsilon) < \varepsilon.$$

Hence, we can choose a sequence η_n adapted to the $\sigma-$fields $\mathcal{F}_{n,1}$ so that (2.15) is satisfied. Furthermore, (2.20) is obviously true, then we can obtain the (2.21) from the theorem. □

2.2.3 *Extension of martingale invariance principle*

As one of main conventional tools, the classical martingale invariance principle is widely used in statistics, econometrics and other fields. In many applications, however, the condition (2.15) seems to be too restrictive. For example, consider

$$M_{n,i} = c_n \sum_{m=1}^{i} g(X_{n,m})\varepsilon_{m+1}, \tag{2.44}$$

where g is a real integrable function on $\mathbb{R}$, ε_m is a stationary sequence with mean 0. For every $n > 0$, $\{X_{n,m}\}_{m=1}^{k_n}$ is a random walk. Let $\mathcal{F}_{n,m} = \sigma(X_{n,1}, \cdots, X_{n,m}; \varepsilon_1, \cdots, \varepsilon_m)$ and assume $\mathbb{E}(\varepsilon_{m+1}^2 | \mathcal{F}_{n,m}) = 1$. Obviously, $\{M_{n,i}, \mathcal{F}_{n,i}\}$ is a martingale array with conditional variance

$$U_n^2 = c_n^2 \sum_{m=1}^{k_n} g^2(X_{n,m}).$$

Unfortunately, in the statistics and econometric, (2.15) can not be obtained usually, thus the asymptotic of M_{n,k_n} can not be obtained by the classical martingale invariance principle. However, under certain conditions on $X_{n,m}$ such that $X_{n,[n\cdot]} \Rightarrow W(\cdot)$ on $\mathbb{D}[0,1]$, where W is a standard Brownian motion, we may prove that the variance

$$V_n^2 \xrightarrow{d} \eta, \tag{2.45}$$

where η is a random variable. The new question is that whether (2.15) can be replaced by (2.45) in the proof of martingale invariance principle. To deal with this problem, *Wang (2012)* extended the classical martingale central limit theorem to more general case. Under the condition of convergence in distribution of conditional variance, martingale central limit theorem can be obtained.

Assume $\{\varepsilon_{n,m}, \mathcal{F}_{n,m}\}_{1 \le m \le n}$ forms a sequence of martingale differences, Let $\{\eta_{n,m}\}$ and $\{\xi_{n,j}\}$ be two sequences of random variables, and f_n be a real function on $\mathbb{R}^\infty$. Specify

$$X_{n,m} = f_n(\varepsilon_{n,1}, \cdots, \varepsilon_{n,m}; \eta_{n,1}, \cdots, \eta_{n,k}; \xi_{n,1}, \xi_{n,2}, \cdots),$$

and for each $n \ge 1$, the $\mathcal{F}_{n,k}$ is a filtration. Now, we present the extended martingale invariance principle.

Theorem 2.5. *(Extended martingale invariance principle)*

Assume $\{(\eta_{n,m}, \varepsilon_{n,m}), \mathcal{F}_{n,m}\}$ forms a sequence of martingale differences satisfying that

$$\max_{k \le m \le n} |\mathbb{E}(\eta_{n,m+1}^2 | \mathcal{F}_{n,m}) - 1| \to 0, \quad \max_{k \le m \le n} |\mathbb{E}(\varepsilon_{n,m+1}^2 | \mathcal{F}_{n,m}) - 1| \to 0 \ a.s. \tag{2.46}$$

as $n \to \infty$ first and then $k \to \infty$, and as $A \to \infty$,

$$\max_{1 \le m \le n} [\mathbb{E}(\eta_{n,m+1}^2 1_{\{|\eta_{n,m+1}| \ge A\}} | \mathcal{F}_{n,m}) + \mathbb{E}(\varepsilon_{n,m+1}^2 1_{\{|\varepsilon_{n,m+1}| \ge A\}} | \mathcal{F}_{n,m})] \to 0 \ a.s. \tag{2.47}$$

Furthermore, assume

$$\max_{1\le m\le n}|X_{n,m}| = o_{\mathbb{P}}(1), \tag{2.48}$$

$$\frac{1}{\sqrt{n}}\sum_{m=1}^{n}|X_{n,m}||\mathbb{E}(\eta_{n,m+1}\varepsilon_{n,m+1}|\mathcal{F}_{n,m})| = o_{\mathbb{P}}(1) \tag{2.49}$$

and

$$\{\frac{1}{\sqrt{n}}\sum_{m=1}^{[n\cdot]}\eta_{n,m+1}, \sum_{m=1}^{n}X_{n,m}^2\} \Rightarrow \{W, g^2(W)\} \tag{2.50}$$

on $\mathbb{D}[0,1]\times\mathbb{R}$, *where* W *is standard Brownian motion, and* $g^2(W)$ *is an a.s. finite functional of* W. *Let* $S_n(t) = \sum_{m=1}^{[nt]}X_{n,m}\varepsilon_{n,m+1}$, $G_n^2 = \sum_{m=1}^{n}X_{n,m}^2$. *Then*

$$\{S_n, G_n^2\} \Rightarrow \{g(W)B, g^2(W)\} \tag{2.51}$$

where B *is a standard Brownian motion independent of* W.

Before proving Theorem 2.5, we first present two useful lemmas. Choose some $\lambda > 0$, such that

$$\mathbb{P}(g^2(W) > \lambda) = 0, \tag{2.52}$$

and λ is the continuous point of the distribution function of $g^2(W)$. Let

$$X_{n,m}^* = X_{n,m}1_{\{\sum_{i=1}^{m}X_{n,i}^2\le 2\lambda\}}, S_n^* = \sum_{m=1}^{n}X_{n,m}^*\varepsilon_{n,m+1} \text{ and } G_n^{*2} = \sum_{m=1}^{n}X_{n,m}^{*2}.$$

For $\alpha, \beta_k \in \mathbb{R}$, $0 = u_0 < u_1 < \cdots < u_N = 1$, define

$$V_n = \sum_{m=1}^{N}\beta_m[W_n(u_m) - W_n(u_{m-1})]$$

where $W_n(t) = \frac{1}{\sqrt{n}}\sum_{m=1}^{[nt]}\eta_{n,m+1}$, and

$$\Gamma_n = \sum_{m=1}^{n}\mathbb{E}\{[\exp(\mathrm{i}\beta_m^* + \mathrm{i}\alpha X_{n,m}^*\varepsilon_{n,m+1}) - 1|\mathcal{F}_{n,m}]\},$$

where $\beta_m^* = \frac{\beta_j}{\sqrt{n}}$ when $[nu_{j-1}] < m \le [nu_j]$ for $j = 1, \cdots, N$.

Lemma 2.3. *For any* $\alpha, \beta_k \in \mathbb{R}$ *and* $0 = u_0 < u_1 < \cdots < u_N = 1$, $\exp(|\Gamma_n|)$ *is uniformly integrable and*

$$(V_n, G_n^{*2}, \Gamma_n) \xrightarrow{d} (V, g^2(W), \Gamma) \tag{2.53}$$

where $V = \sum_{m=1}^{N}\beta_m^2(W(u_m) - W(u_{m-1}))$ *and* $\Gamma = -\frac{1}{2}\sum_{m=1}^{N}\beta_m^2(u_m - u_{m-1}) - \frac{1}{2}\alpha^2 g^2(W)$.

Proof. Recall that $\max_{1\leq m\leq n}|X_{n,m}| = o_{\mathbb{P}}(1)$. There exists a sequence $A_n > 0$, such that

$$A_n \to \infty, \quad \frac{A_n}{\sqrt{n}} \to 0, \quad A_n \max_{1\leq m\leq n}|X_{n,m}| = o_{\mathbb{P}}(1). \tag{2.54}$$

Let $Y_{nm} = \beta_m^* \eta_{n,m+1} + \alpha X_{n,m}\varepsilon_{n,m+1}$, $\Omega_{nm} = \{|\eta_{n,m+1}| \leq A_n, |\varepsilon_{n,m+1}| \leq A_n\}$.

Noting $|\beta_m^*| \leq C/\sqrt{n}$, obviously, we have

$$\mathbb{E}[|Y_{nm}|^3 1_{\{\Omega_{nm}\}}|\mathcal{F}_{n,m}] \leq C(n^{-1} + X_{n,m}^{*2})R_{1nm}, \tag{2.55}$$

$$\mathbb{E}[Y_{nm}^2 1_{\{\Omega_{nm}^c\}}|\mathcal{F}_{n,m}] \leq C(n^{-1} + X_{n,m}^{*2})R_{2nm}, \tag{2.56}$$

where

$$R_{1nm} = (A_n n^{-1/2} + A_n \max_{1\leq k\leq n}|X_{n,k}|)(\mathbb{E}[\eta_{n,m+1}^2|\mathcal{F}_{n,m}] + \mathbb{E}[\varepsilon_{n,m+1}^2|\mathcal{F}_{n,m}]),$$

$$R_{2nm} = \mathbb{E}[(\eta_{n,m+1}^2 + \varepsilon_{n,m+1}^2)1_{\{\Omega_{nm}^c\}}|\mathcal{F}_{n,m}].$$

By the Taylor expansion

$$\begin{aligned}
&|\mathbb{E}[\exp(\mathrm{i}Y_{nm}) - 1 - \mathrm{i}Y_{nm}|\mathcal{F}_{n,m}]| \\
\leq& -\frac{1}{2}\mathbb{E}[Y_{nm}^2 1_{\{\Omega_{nm}\}}|\mathcal{F}_{n,m}] + |\mathbb{E}[(\exp(\mathrm{i}Y_{nm}) - 1 - \mathrm{i}Y_{nm})1_{\{\Omega_{nm}^c\}}|\mathcal{F}_{n,m}]| \\
&+|\mathbb{E}[(\exp(\mathrm{i}Y_{nm}) - 1 - \mathrm{i}Y_{nm} + \frac{1}{2}Y_{nm}^2)1_{\{\Omega_{nm}\}}|\mathcal{F}_{n,m}]| \\
\leq& -\frac{1}{2}\mathbb{E}[Y_{nm}^2|\mathcal{F}_{n,m}] + \mathbb{E}[Y_{nm}^2 1_{\{\Omega_{nm}^c\}}|\mathcal{F}_{n,m}] + \mathbb{E}[|Y_{nm}|^3 1_{\{\Omega_{nm}\}}|\mathcal{F}_{n,m}] \\
\leq& -\frac{1}{2}\mathbb{E}[Y_{nm}^2|\mathcal{F}_{n,m}] + C(n^{-1} + X_{n,m}^{*2})(R_{1nm} + R_{2nm}).
\end{aligned}$$

Due to (2.47) and (2.54),

$$\begin{aligned}
&\sum_{m=1}^{n}(\mathbb{E}[Y_{nm}^2 1_{\{\Omega_{nm}^c\}}|\mathcal{F}_{n,m}] + \mathbb{E}[|Y_{nm}|^3 1_{\{\Omega_{nm}\}}|\mathcal{F}_{n,m}]) \\
&\leq C\max_{1\leq m\leq n}(R_{1nm} + R_{2nm})\sum_{m=1}^{n}(n^{-1} + X_{n,m}^{*2}) = o_{\mathbb{P}}(1).
\end{aligned}$$

Thus

$$\begin{aligned}
\Gamma_n &= \sum_{m=1}^{n}\mathbb{E}[\exp(iY_{nm}) - 1 - iY_{nm}|\mathcal{F}_{n,m}] \\
&= -\frac{1}{2}\sum_{m=1}^{n}\mathbb{E}[(\beta_m^*\eta_{n,m+1} + \alpha X_{n,m}^*\varepsilon_{m+1})^2|\mathcal{F}_{n,m}] + o_{\mathbb{P}}(1) \\
&= -\frac{1}{2}\sum_{m=1}^{n}\beta_m^{*2}\mathbb{E}[\eta_{n,m+1}^2|\mathcal{F}_{n,m}] - \frac{\alpha^2}{2}\sum_{m=1}^{n}\mathbb{E}[X_{n,m}^{*2}\varepsilon_{m+1}^2|\mathcal{F}_{n,m}] \\
&\quad -\sum_{m=1}^{n}\beta_m^*\alpha X_{n,m}^*\mathbb{E}[\eta_{n,m+1}\varepsilon_{m+1}|\mathcal{F}_{n,m}] + o_{\mathbb{P}}(1)
\end{aligned}$$

$$= -\frac{1}{2n}\sum_{m=1}^{N}\beta_m^2([nu_m]-[nu_{m-1}]) - \frac{\alpha^2}{2}\sum_{m=1}^{n}X_{n,m}^{*2} + o_{\mathbb{P}}(1)$$
$$= -\frac{1}{2}\sum_{m=1}^{N}\beta_m^2(u_m-u_{m-1}) - \frac{\alpha^2}{2}\sum_{m=1}^{n}X_{n,m}^{*2} + o_{\mathbb{P}}(1)$$

where we have used the fact

$$|\sum_{m=1}^{n}\beta_m^*\alpha X_{n,m}^*\mathbb{E}[\eta_{n,m+1}\varepsilon_{m+1}|\mathcal{F}_{n,m}]|$$
$$\leq \frac{C}{\sqrt{n}}\sum_{m=1}^{n}|X_{n,m}||\mathbb{E}(\eta_{n,m+1}\varepsilon_{n,m+1}|\mathcal{F}_{n,m})| = o_{\mathbb{P}}(1)$$

by (2.49). So, for any $\alpha_i\in\mathbb{R}, i=1,2,3$, we have

$$\alpha_1V_n+\alpha_2G_n^{*2}+\alpha_3\Gamma_n$$
$$= \alpha_1V_n + (\alpha_2-\frac{\alpha^2\alpha_3}{2})G_n^{*2} - \frac{\alpha_3}{2}\sum_{m=1}^{N}\beta_m^2(u_m-u_{m-1}) + o_{\mathbb{P}}(1)$$
$$\xrightarrow{d} \alpha_1V + (\alpha_2-\frac{\alpha^2\alpha_3}{2})g^2(W) - \frac{\alpha_3}{2}\sum_{m=1}^{N}\beta_m^2(u_m-u_{m-1})$$
$$= \alpha_1V+\alpha_2g^2(W)+\alpha_3\Gamma.$$

Then (2.53) holds. Furthermore, noting that

$$\mathbb{E}[\eta_{n,m+1}^2+\varepsilon_{n,m+1}^2|\mathcal{F}_{n,m}]\leq C$$

and

$$|\mathbb{E}[\exp(iY_{nm})-1|\mathcal{F}_{n,m}]| \leq \frac{1}{2}\mathbb{E}[Y_{nm}^2|\mathcal{F}_{n,m}] \leq C(n^{-1}+X_{n,m}^2),$$

we have

$$|\Gamma_n|\leq C(1+\sum_{m=1}^{n}X_{n,m}^2)\leq C(1+2\lambda).$$

The uniformly integrability of $\exp(|\Gamma_n|)$ is easily obtained. □

Lemma 2.4. *For any $\alpha,\beta_k\in\mathbb{R}$ and $0=u_0<u_1<\cdots<u_N=1$, we have*

$$I_n := \mathbb{E}|\mathbb{E}[\exp(\mathrm{i}\alpha S_n^* + \mathrm{i}V_n - \Gamma_n)|\mathcal{F}_{n,1}] - 1| = o(1). \tag{2.57}$$

Proof. Recall that $Y_{nm} = \beta_m^*\eta_{n,m+1} + \alpha X_{n,m}\varepsilon_{n,m+1}$, and write $Z_{nm} = \mathbb{E}[\exp(iY_{nm})-1|\mathcal{F}_{n,m}]$. Due to the definitions of $W_n(t)$ and β_k^*, we have

$$I_n = \mathbb{E}|\mathbb{E}[\exp(\mathrm{i}\sum_{k=1}^{n}Y_{nk} - \sum_{k=1}^{n}Z_{nk})|\mathcal{F}_{n,1}] - 1|$$
$$\leq \mathbb{E}|\mathbb{E}[\exp(\mathrm{i}\sum_{k=1}^{n-1}Y_{nk} - \sum_{k=1}^{n}Z_{nk})\exp(\mathrm{i}Y_{nn}) - \exp(Z_{nn}))|\mathcal{F}_{n,1}]|$$

$$+\mathbb{E}|\mathbb{E}[\exp(\mathrm{i}\sum_{k=1}^{n-1} Y_{nk} - \sum_{k=1}^{n-1} Z_{nk})|\mathcal{F}_{n,1}] - 1|.$$

By induction

$$\begin{aligned} I_n &\leq \sum_{m=2}^{n} \mathbb{E}|\mathbb{E}[\exp(\mathrm{i}\sum_{k=1}^{m-1} Y_{nk} - \sum_{k=1}^{m} Z_{nk})(\exp(\mathrm{i}Y_{nm}) - \exp(Z_{nm}))|\mathcal{F}_{n,1}]| \\ &\quad +\mathbb{E}|\mathbb{E}[\exp(\mathrm{i}Y_{n1} - Z_{n1})|\mathcal{F}_{n,1}] - 1| \\ &=: I_{1n} + I_{2n}. \end{aligned}$$

Where, by the Jensen inequality,

$$\begin{aligned} I_{2n} &= |\mathbb{E}[\exp(\mathrm{i}Y_{n1} - Z_{n1})|\mathcal{F}_{n,1}] - 1| \\ &= |\mathbb{E}[\exp(iY_{n1} - \mathbb{E}[\exp(\mathrm{i}Y_{n1}) - 1|\mathcal{F}_{n,1}])|\mathcal{F}_{n,1}] - 1| \\ &\leq |\exp(-\mathbb{E}[\exp(\mathrm{i}Y_{n1}) - 1|\mathcal{F}_{n,1}]) - 1| + |\mathbb{E}[\exp(\mathrm{i}Y_{n1}) - 1|\mathcal{F}_{n,1}]| \\ &\leq C(n^{-1} + X_{n,1}^{*2}) \exp(C(n^{-1} + X_{n,1}^{*2})) + o_{\mathbb{P}}(1), \end{aligned}$$

and the uniformly integrability of $|\mathbb{E}[(\exp(\mathrm{i}Y_{n1} - Z_{n1})|\mathcal{F}_{n,1}] - 1|$ implies $I_{2n} \to 0$ as $n \to \infty$.

Define

$$\begin{aligned} u_{n,m} &= \exp(\mathrm{i}\sum_{k=1}^{m-1} Y_{nk} - \sum_{k=1}^{m} \mathbb{E}[\exp(\mathrm{i}Y_{nk}) - 1|\mathcal{F}_{n,k}]), \\ v_{n,m} &= \exp(\mathrm{i}Y_{nm}) - \exp(\mathbb{E}[\exp(\mathrm{i}Y_{nm}) - 1|\mathcal{F}_{n,m}]). \end{aligned}$$

$\sum_{k=1}^{n} X_{n,k}^{*2} \leq 2\lambda$ implies

$$|u_{n,m}| \leq \exp(\sum_{k=1}^{m} |\mathbb{E}[\exp(\mathrm{i}Y_{nk}-1)|\mathcal{F}_{n,k}]|) \leq \exp(C\sum_{k=1}^{m}(n^{-1}+X_{n,k}^{*2})) \leq \exp(C(1+2\lambda)). \tag{2.58}$$

Since $|\exp(x) - 1 - x| \leq |x|^{(2+\delta)/2} \exp(|x|)$ for some $\delta > 0$ and any $x \in \mathbb{R}$, we have

$$\begin{aligned} |\mathbb{E}(v_{n,m}|\mathcal{F}_{n,m})| &= |\mathbb{E}[\exp(\mathrm{i}Y_{nm}) - 1|\mathcal{F}_{n,m}] + 1 - \exp(\mathbb{E}[\exp(\mathrm{i}Y_{nm}) - 1|\mathcal{F}_{n,m}])| \\ &\leq |\mathbb{E}[\exp(\mathrm{i}Y_{nm}) - 1|\mathcal{F}_{n,m}]|^{(2+\delta)/2} \exp(|\mathbb{E}[\exp(\mathrm{i}Y_{nm}) - 1|\mathcal{F}_{n,m}]|) \\ &\leq C \exp(C(1 + 2\lambda))(n^{-1} + X_{n,m}^{*2})^{(2+\delta)/2}. \end{aligned}$$

By recalling $\mathcal{F}_{n,m} \subseteq \mathcal{F}_{n,m+1}$ for any $n \geq m \geq 1$, and $n \geq 1$, it is readily seen that

$$\begin{aligned} I_{1n} &\leq \sum_{m=2}^{n} \mathbb{E}[|u_{n,m}||\mathbb{E}[v_{n,m}|\mathcal{F}_{n,m}]|] \\ &\leq C \exp(2C(1 + 2\lambda))\mathbb{E}[\sum_{m=2}^{n}(n^{-1} + X_{n,m}^{*2})^{(2+\delta)/2}] \to 0, \end{aligned}$$

since

$$\begin{aligned} &\sum_{m=2}^{n}(n^{-1} + X_{n,m}^{*2})^{(2+\delta)/2} \\ &\leq (n^{-\delta/2} + \max_{1\leq m\leq n} |X_{n,m}^{*}|^{\delta})(1 + \sum_{m=1}^{n} X_{n,m}^{*2}) = o_{\mathbb{P}}(1) \end{aligned}$$

and $\sum_{m=2}^{n}(n^{-1} + X_{n,m}^{*2})^{(2+\delta)/2}$ is uniformly integrable, (2.57) is obtained. □

The proof of Theorem 2.5. According to Theorem 2.4, we need to prove convergence of finite dimensional distributions and tightness for $\{S_n, G_n^2\}$.

For convergence of finite dimensional distributions, we need to prove that

$$\lim_{n\to\infty} |\mathbb{E}\exp[\sum_{k=1}^{p} z_k(S_n(t_k) - S_n(t_{k-1})) + \mathrm{i}sG_n^2]$$
$$- \mathbb{E}\exp[-\frac{1}{2}g^2(W)\sum_{k=1}^{p} z_k^2(t_k - t_{k-1}) + \mathrm{i}sg^2(W)]| = 0 \tag{2.59}$$

for all sequence $0 = t_0 < t_1 < \cdots < t_p \le 1$ and real numbers $z_1, z_2, \cdots, z_p$, s. For facilitating the proof, we only need to prove that

$$\lim_{n\to\infty} |\mathbb{E}\exp[\mathrm{i}\alpha S_n + \mathrm{i}\beta G_n^2] - \mathbb{E}\exp[(-\frac{1}{2}\alpha + \mathrm{i}\beta)g^2(W)]| = 0, \tag{2.60}$$

where $S_n = S_n(1)$, $\alpha, \beta \in \mathbb{R}$.

On the set $\{G_n^2 \le 2\lambda\}$, $S_n = S_n^*$ and $G_n^2 = G_n^{*2}$. By (2.50),

$$\lim_{n\to\infty} \mathbb{P}(G_n^2 > 2\lambda) = 0.$$

It is easy to see that, for all $\alpha, \beta \in \mathbb{R}$,

$$\mathbb{E}|\exp(\mathrm{i}\alpha S_n^* + \mathrm{i}\beta G_n^{*2}) - \exp(\mathrm{i}\alpha S_n + \mathrm{i}\beta G_n^2)| \le 2\mathbb{P}(G_n^2 > 2\lambda) \to 0. \tag{2.61}$$

Hence, (2.60) holds true iff

$$\lim_{n\to\infty} |\mathbb{E}\exp[\mathrm{i}\alpha S_n^* + \mathrm{i}\beta G_n^{*2}] - \mathbb{E}\exp[(-\frac{1}{2}\alpha + \mathrm{i}\beta)g^2(W)]| = 0. \tag{2.62}$$

Since $\mathbb{E}S_n^{*2} \le C$, $\mathbb{E}G_n^{*2} \le C\lambda$ and (2.53),

$$\{S_n^*, G_n^{*2}, V_n, \Gamma_n\}_{n\ge 1}$$

is tight. Hence, for each subsequence $\{n'\} \subset \{n\}$, there exists a further subsequence $\{n''\} \subset \{n'\}$ such that

$$\{S_{n''}^*, G_{n''}^{*2}, V_{n''}, \Gamma_{n''}\} \xrightarrow{d} \{S, g^2(W), V, \Gamma\},$$

where S is a limiting random variable of $S_{n''}^*$. By Lemma 2.3, we have

$$\mathbb{E}[\exp(\mathrm{i}\alpha S_{n''}^* + \mathrm{i}V_{n''} - \Gamma_{n''})] \to \mathbb{E}[\exp(\mathrm{i}\alpha S + \mathrm{i}V - \Gamma)].$$

This, together with Lemma 2.4, yields that $\mathbb{E}[\exp(\mathrm{i}\alpha S + \mathrm{i}V - \Gamma)] = 1$. Hence, by noting $\mathbb{E}[\exp(\mathrm{i}V)] = \exp(-\frac{1}{2}\sum_{k=1}^{N}\beta_k(u_k - u_{k-1}))$, we obtain

$$\mathbb{E}[\exp(\mathrm{i}\alpha S + \mathrm{i}V + \frac{1}{2}\alpha^2 g^2(W))] = \exp(-\frac{1}{2}\sum_{k=1}^{N}\beta_k(u_k - u_{k-1})).$$

Furthermore, using the Weierstrass theorem on approximation of continuous functions by trigonometric polynomials, we have

$$\mathbb{E}[\exp(\mathrm{i}\alpha S + \frac{1}{2}\alpha^2 g^2(W)) - 1]G[W(u_0), W(u_1), \cdots, W(u_N)] = 0 \tag{2.63}$$

for any $0 = u_0 < u_1 < \cdots < u_N = 1$, where $G(y_0, \cdots, y_N)$ is an arbitrary continuous function with compact support. Write $\mathcal{G} = \sigma(W(s), 0 \le s \le 1)$. (2.63) means that

$$\mathbb{E}[\exp(\mathrm{i}\alpha S + \frac{1}{2}\alpha^2 g^2(W)) - 1|\mathcal{G}] = 0.$$

Consequently, for all $\alpha, \beta \in \mathbb{R}$,

$$\mathbb{E}[\exp(\mathrm{i}\alpha S + \mathrm{i}\beta g^2(W))|\mathcal{G}] = \exp((-\frac{1}{2}\alpha^2 + \mathrm{i}\beta)g^2(W)).$$

Hence, for each $\{n'\} \subset \{n\}$, there exists a subsequence $\{n''\} \subset \{n'\}$ such that

$$\lim_{n''\to\infty} \mathbb{E}\exp[\mathrm{i}\alpha S^*_{n''} + \mathrm{i}\beta G^{*2}_{n''}] = \mathbb{E}[\exp(\mathrm{i}\alpha S + \mathrm{i}\beta g^2(W))] = \mathbb{E}[\exp((-\frac{1}{2}\alpha^2 + \mathrm{i}\beta)g^2(W))].$$

(2.60) is established as the limitation does not depend on the choice of the subsequences, and also completes the proof in the special case where (2.52) holds.

It remains to remove the boundedness condition (2.52). To this end, for given $\varepsilon > 0$, choose a continuous point λ of distribution function of $g^2(W)$ such that $\mathbb{P}(g^2(W) > \lambda) \le \varepsilon$. Let $g^2_\lambda(W) = g^2(W)1_{\{g^2(W)\le\lambda\}} + \lambda 1_{\{g^2(W)>\lambda\}}$, $X'_{n,k} = X_{n,k}1_{\{\sum_{i=1}^k X^2_{n,i}\le\lambda\}}$,

$$S'_n = \sum_{k=1}^n X'_{n,k}\varepsilon_{n,k+1}, \quad G'^2_n = \sum_{k=1}^n X'^2_{n,k}.$$

Now $\widetilde{G}^2_n - \max_{1\le k\le n} X^2_{n,k} \le G'^2_n \le \widetilde{G}^2_n$, where

$$\widetilde{G}^2_n = G^2_n 1_{\{G^2_n\le\lambda\}} + \lambda 1_{\{G^2_n>\lambda\}}.$$

Since (2.49) and (2.50),

$$\frac{1}{\sqrt{n}}\sum_{k=1}^n |X'_{n,k}||\mathbb{E}(\eta_{n,k+1}\varepsilon_{n,k+1}|\mathcal{F}_{n,k})| \le \frac{1}{\sqrt{n}}\sum_{k=1}^n |X_{n,k}||\mathbb{E}(\eta_{n,k+1}\varepsilon_{n,k+1}|\mathcal{F}_{n,k})| = o_{\mathbb{P}}(1)$$

and $G'^2_n \xrightarrow{d} g^2_\lambda(W)$ since $\widetilde{G}^2_n \xrightarrow{d} g^2_\lambda(W)$ by the continuous mapping theorem. Note that $g^2_\lambda(W)$ is a.s. bounded, so

$$\begin{aligned}
&\lim_{n\to\infty} |\mathbb{E}[\exp(\mathrm{i}\alpha S_n + \mathrm{i}\beta G^2_n)] - \mathbb{E}[\exp((-\frac{\alpha^2}{2} + \mathrm{i}\beta)g^2(W))]| \\
&\le \lim_{n\to\infty} |\mathbb{E}[\exp(\mathrm{i}\alpha S'_n + \mathrm{i}\beta G'^2_n)] - \mathbb{E}[\exp((-\frac{\alpha^2}{2} + \mathrm{i}\beta)g^2_\lambda(W))]| \\
&\quad +2\lim_{n\to\infty} \mathbb{P}(G^2_n > \lambda) + 2\mathbb{P}(g^2(W) > \lambda) \\
&\le 4\varepsilon
\end{aligned}$$

which implies (2.60).

Consider tightness of $\{S_n, G^2_n\}$. We only need to prove the tightness of $\{S_n\}$, since weak convergence of $\{G_n\}$ by (2.50). In fact, for every $\varepsilon > 0$,

$$\begin{aligned}
&\sup_{0<t-s\le\delta} \mathbb{P}(|\sum_{k=[ns]+1}^{[nt]} X_{n,k}\varepsilon_{n,k+1}| > \varepsilon) \\
&\le \varepsilon^{-2} \sup_{0<t-s\le\delta} \sum_{k=[ns]+1}^{[nt]} \mathbb{E}X^2_{n,k+1},
\end{aligned}$$

which implies tightness of $\{S_n\}$ in $\mathbb{D}[0,1]$. □

2.3 Convergence of Poisson point processes

In the previous section, Donsker type invariance principles are studied under second moment conditions (such as conditions (2.14), (2.46), (2.47)). When variances do not exist, weak convergence of underlying processes may be changed. For example, the data, appearing in the fields of data network, financial returns and reinsurance, usually are heavy tailed. Roughly speaking, a random variable X is called *heavy tailed* if there exists a positive parameter $\alpha > 0$ such that

$$\mathbb{P}[X > x] \sim x^{-\alpha} \quad \text{as } x \to \infty.$$

Here and elsewhere that we use

$$g(x) \sim h(x) \quad \text{as } x \to \infty$$

as shorthand for

$$\lim_{x\to\infty} \frac{g(x)}{h(x)} = 1.$$

Heavy-tail analysis is an important branch of probability theory. Its mathematical tool is based on point processes. Our interest in this section is the weak convergence in the context of heavy-tail analysis.

We firstly present some preliminaries which are used in the section.

1. Regular variation

In our study of heavy-tail analysis, we assume that the tail probability distribution of heavy tail random variable is regular varying function. Roughly speaking, regularly varying functions are those functions which behave asymptotically like power functions. Its mathematical definition is following:

Definition 2.1. A measurable function $U : \mathbb{R}_+ \to \mathbb{R}_+$ is regularly varying at ∞ with index $\rho \in \mathbb{R}$ (written $U \in RV_\rho$) if for any $x > 0$,

$$\lim_{t\to\infty} \frac{U(tx)}{U(t)} = x^{\rho}.$$

If $\rho = 0$, we call U slowly varying.

Example 2.1. The extreme-value distribution

$$\Phi_\alpha(x) = \exp(-x^{-\alpha}), \quad x > 0,$$

its tail satisfies

$$1 - \Phi_\alpha(x) \sim x^{-\alpha} \quad \text{as } x \to \infty.$$

Example 2.2. The Cauchy distribution F has density function $f(x) = \frac{1}{\pi(1+x^2)}$, and its tail satisfies

$$1 - F(x) \sim (\pi x)^{-1} \quad \text{as } x \to \infty.$$

Example 2.3.

A random variable X is said to have a $\alpha-$*stable distribution* if for any positive numbers A and B, there are a positive number C and a real number D such that

$$AX_1 + BX_2 \stackrel{d}{=} CX + D,$$

where X_1 and X_2 are independent copies of X, and there exists a number $\alpha \in (0, 2]$ such that $C^\alpha = A^\alpha + B^\alpha$. When $0 < \alpha < 2$, $\alpha-$stable distribution G satisfies

$$1 - G(x) \sim cx^{-\alpha} \quad \text{as } x \to \infty.$$

Regular variation of distribution tails can be reformulated by vague convergence. This will be useful in the study of weak convergence.

Proposition 2.1. *Suppose that ξ is a nonnegative random variable with distribution function F. Set $\overline{F} = 1 - F$. The following statements are equivalent:*

(i) $\overline{F} \in RV_{-\alpha}$, $\alpha > 0$.

(ii) There exists a sequence $\{b_n\}$ with $b_n \to \infty$ such that

$$\lim_{n\to\infty} n\overline{F}(b_n x) = x^{-\alpha}, \ x > 0.$$

(iii) There exists a sequence $\{b_n\}$ with $b_n \to \infty$ such that

$$n\mathbb{P}[\frac{\xi}{b_n} \in \cdot] \stackrel{v}{\to} \nu_\alpha(\cdot)$$

in $\mathbb{M}_+(0, \infty]$, where $\nu_\alpha((x, \infty]) = x^{-\alpha}$.

Proof.

(i)$\Rightarrow$ (ii). Let b_n be a quantile of F, (ii) is easily obtained.

(ii)$\Rightarrow$ (i). Define

$$n(t) = \inf\{n : \ b_{n+1} > t\}.$$

Note that

$$b_{n(t)} \le t < b_{n(t)+1}.$$

By monotonicity, we have

$$\frac{n(t)+1}{n(t)} \cdot \frac{n(t)\overline{F}(b_{n(t)}x)}{(n(t)+1)\overline{F}(b_{n(t)+1})} \le \frac{\overline{F}(tx)}{\overline{F}(t)} \le \frac{n(t)}{n(t)+1} \cdot \frac{(n(t)+1)\overline{F}(b_{n(t)+1}x)}{n(t)\overline{F}(b_{n(t)})},$$

and so

$$\lim_{t\to\infty} \frac{\overline{F}(tx)}{\overline{F}(t)} = x^{-\alpha}$$

holds.

(ii)$\Rightarrow$ (iii). Let $\mathbb{C}_K^+((0, \infty])$ be the space of continuous functions which have compact supports. For $f \in \mathbb{C}_K^+((0, \infty])$, the support of f is contained in $(\delta, \infty]$ for some $\delta > 0$. From (ii),

$$n\mathbb{P}[\frac{\xi}{b_n} > x] \to x^{-\alpha} = \nu_\alpha((x, \infty]) \text{ for any } x > 0.$$

On $(\delta, \infty]$, define

$$P_n(\cdot) = \frac{\mathbb{P}[\frac{\xi}{b_n} \in \cdot]}{\mathbb{P}[\frac{\xi}{b_n} > \delta]}$$

and

$$P(\cdot) = \frac{\nu_\alpha(\cdot)}{\nu_\alpha((\delta, \infty])},$$

which are probability measures on $(\delta, \infty]$. Then for $y \in (\delta, \infty]$,

$$P_n((y, \infty]) \to P((y, \infty]).$$

In $\mathbb{R}$, convergence of distribution functions is equivalent to weak convergence, so for bounded continuous function f on $(\delta, \infty]$, we have

$$P_n(f) \to P(f)$$

which means that

$$n\mathbb{E}f(\frac{\xi}{b_n}) \to \nu_\alpha(f).$$

(iii) is obtained.

(iii)$\Rightarrow$ (ii). From (iii), since $(x, \infty]$ is a relatively compact set, and

$$\nu_\alpha(\partial(x, \infty]) = 0$$

so we have

$$n\mathbb{P}[\frac{\xi}{b_n} > x] \to x^{-\alpha} = \nu_\alpha((x, \infty]) \text{ for any } x > 0.$$

(ii) is obtained. □

2. Point process

Point process is a well studied object in probability theory, and the point process method is a powerful tool in statistics. In the study of weak convergence under heavy-tailed assumption, point process serves as a tool, bridging gap between heavy-tail analysis and weak convergence. Definition of point process is based on the point measure.

Definition 2.2. The random element $\mathcal{N} : (\Omega, \mathcal{F}) \to (\mathbb{M}_p(\mathbb{S}), \mathcal{M}_p(\mathbb{S}))$ is called as point process with state space $\mathbb{S}$, where $\mathcal{M}_p(\mathbb{S})$ is the Borel $\sigma-$filed of $\mathbb{M}_p(\mathbb{S})$ generated by the open sets in $\mathbb{M}_p(\mathbb{S})$.

The set of Borel subsets in $\mathbb{S}$ is denoted by $\mathcal{S}$.

Definition 2.3. $\mathbb{N}$ is a *Poisson random measure* with mean measure μ ($PRM(\mu)$), if

(1) for $A \in \mathcal{S}$, $\mathbb{P}[\mathbb{N}(A) = k] = \frac{\exp(-\mu(A))(\mu(A))^k}{k!}$ if $\mu(A) < \infty$, and $\mathbb{P}[\mathbb{N}(A) = k] = 0$ if $\mu(A) = \infty$;

(2) for disjoint subsets of $\mathbb{S}$ in $\mathcal{S}$, $A_1, \cdots, A_k$, $\mathbb{N}(A_1), \cdots, \mathbb{N}(A_k)$ are independent random variables.

2.3.1 *Functional limit theorem for independent heavy-tailed sequence*

The limiting process in this section is pure-jump Lévy process. The distribution of a Lévy process $Z(\cdot)$ is characterized by its characteristic Lévy triple, that is, the characteristic triple of the infinitely divisible distribution of $Z(1)$. The characteristic function of $Z(1)$ and the *characteristic Lévy triple* (b, c, ν) of Z are related in the following way:

$$\mathbb{E}[\exp(iuZ(1))] = \exp(ibu - \frac{1}{2}cu^2 + \int_{\mathbb{R}} (e^{iux} - 1 - iux1_{\{|x|\le 1\}}(x))\nu(dx))$$

for $u \in \mathbb{R}$, where $b \in \mathbb{R}$, $c \ge 0$, and ν is a measure on $\mathbb{R}$. When $c = 0$, the Lévy process is a pure-jump process.

Firstly, we present a basic convergence lemma.

Lemma 2.5. *Suppose that for each $n \ge 1$, $\{X_{n,j}, j \ge 1\}$ is a sequence of i.i.d. random elements of $(\mathbb{S}, \mathcal{S})$. Let ξ be PRM(μ) on $\mathbb{M}_p(\mathbb{S})$.*

(i) We have

$$\sum_{j=1}^{n} \delta_{X_{n,j}} \Rightarrow \xi$$

on $\mathbb{M}_+(\mathbb{S})$ iff

$$n\mathbb{P}[X_{n,1} \in \cdot] = \mathbb{E}\left(\sum_{j=1}^{n} \delta_{X_{n,j}}(\cdot)\right) \xrightarrow{v} \mu \tag{2.64}$$

in $\mathbb{M}_+(\mathbb{S})$.

(ii) For a measure $\mu \in \mathbb{M}_+(\mathbb{S})$, we have

$$\frac{1}{a_n}\sum_{j=1}^{n} \delta_{X_{n,j}} \Rightarrow \mu \tag{2.65}$$

on $\mathbb{M}_+(\mathbb{S})$ for some a_n with $0 < a_n \uparrow \infty$ iff

$$\frac{n}{a_n}\mathbb{P}[X_{n,1} \in \cdot] = \mathbb{E}\left(\frac{1}{a_n}\sum_{j=1}^{n} \delta_{X_{n,j}}(\cdot)\right) \xrightarrow{v} \mu \tag{2.66}$$

in $\mathbb{M}_+(\mathbb{S})$.

Proof. (i) For $f \in \mathbb{C}_K^+(\mathbb{S})$,

$$\begin{aligned}\mathbb{E}[\exp(-\sum_{j=1}^{n} \epsilon_{X_{n,j}(f)})] &= \mathbb{E}[\exp(-\sum_{j=1}^{n} f(X_{n,j}))] = (\mathbb{E}\exp(-f(X_{n,1})))^n \\ &= \left(1 - \frac{\mathbb{E}(n(1-\exp(-f(X_{n,1}))))}{n}\right)^n\end{aligned}$$

$$= \left(1 - \frac{\int_{\mathbb{S}}(1 - \exp(-f(x)))n\mathbb{P}[X_{n,1} \in dx]}{n}\right)^n,$$

which converges to

$$\exp\left\{\int_{\mathbb{S}}(1 - \exp(-f(x)))\mu(dx)\right\},$$

the *Laplace functional* of $PRM(\mu)$, iff

$$\int_{\mathbb{S}}(1 - \exp(-f(x)))n\mathbb{P}[X_{n,1} \in dx] \to \int_{\mathbb{S}}(1 - e^{-f(x)})\mu(dx).$$

This last statement is equivalent to vague convergence in (2.64).

(ii) It is enough to prove the Laplace functionals of $\sum_{i=1}^n \epsilon_{X_{n,1}(f)}$ converge. We compute the Laplace functional for the quantity on the left side of (2.65):

$$\begin{aligned}\mathbb{E}\exp(-\frac{\sum_{i=1}^n \epsilon_{X_{n,1}(f)}}{a_n}) &= (\mathbb{E}\exp(-\frac{1}{a_n}f(X_{n,1})))^n \\ &= \left(1 - \frac{\int_{\mathbb{S}}(1 - \exp(-\frac{1}{a_n}f(x)))n\mathbb{P}[X_{n,1} \in dx]}{n}\right)^n,\end{aligned}$$

which converges to $\exp(-\mu(f))$, the Laplace functional of μ, iff

$$\int_{\mathbb{S}}(1 - \exp(-\frac{1}{a_n}f(x)))n\mathbb{P}[X_{n,1} \in dx] \to \mu(f). \tag{2.67}$$

We prove that (2.67) is equivalent to (2.66) as follows: Suppose (2.66) holds. Firstly,

$$\int_{\mathbb{S}}(1 - \exp(-f(x)/a_n))n\mathbb{P}[X_{n,1} \in dx] \leq \int_{\mathbb{S}} f(x)\frac{n}{a_n}\mathbb{P}[X_{n,1} \in dx] \to \mu(f),$$

so

$$\limsup_{n\to\infty} \int_{\mathbb{S}}(1 - \exp(-f(x)/a_n))n\mathbb{P}[X_{n,1} \in dx] \leq \mu(f).$$

Furthermore,

$$\begin{aligned}&\int_{\mathbb{S}}(1 - \exp(-f(x)/a_n))n\mathbb{P}[X_{n,1} \in dx] \\ &\quad\geq \int_{\mathbb{S}} f(x)\frac{n}{a_n}\mathbb{P}[X_{n,1} \in dx] - \int_{\mathbb{S}} \frac{f^2(x)}{2a_n}\frac{n}{a_n}\mathbb{P}[X_{n,1} \in dx] \\ &\quad=: I + II.\end{aligned}$$

Now $I \to \mu(f)$ from (2.66), and

$$II \sim \frac{\mu(f^2)}{2a_n} \to 0$$

since $f^2 \in \mathbb{C}_k^+(\mathbb{S})$ and $a_n \uparrow \infty$. So

$$\liminf_{n\to\infty} \int_{\mathbb{S}}(1 - \exp(-f(x)/a_n))n\mathbb{P}[X_{n,1} \in dx] \geq \mu(f).$$

Hence (2.67) holds.

Conversely, let $f \in \mathbb{C}_+^K(\mathbb{S})$, and suppose that $f \leq 1$. Assuming (2.67), we have

$$f/a_n \geq 1 - \exp(-f/a_n)$$

leading to

$$\liminf_{n\to\infty} \int_{\mathbb{S}} f(x)\frac{n}{a_n}\mathbb{P}[X_{n,1} \in dx] \geq \mu(f)$$

and

$$\frac{f}{a_n} - \frac{f^2}{2a_n^2} \leq 1 - \exp(-f/a_n)$$

leading to

$$\limsup_{n\to\infty} \int_{\mathbb{S}} (\frac{f(x)}{a_n} - \frac{f^2(x)}{2a_n^2})n\mathbb{P}[X_{n,1} \in dx] \leq \mu(f).$$

As before, we may show that

$$\int_{\mathbb{S}} \frac{f^2(x)}{2a_n^2} n\mathbb{P}[X_{n,1} \in dx] \to 0.$$

Hence (2.66) holds. □

Furthermore, we have

Lemma 2.6. *Suppose $\{X, X_1, X_2, \cdots\}$ are i.i.d. random variables. The following statements are equivalent:*
(i)

$$n\mathbb{P}[\frac{X}{b_n} \in \cdot] \xrightarrow{v} \nu$$

in $[-\infty, \infty] \setminus \{0\}$;
(ii)

$$\sum_j \delta_{(\frac{j}{n}, X_j/b_n)} \Rightarrow PRM(\lambda \times \nu) \tag{2.68}$$

in $\mathbb{M}_p([0,\infty) \times [-\infty, \infty] \setminus \{0\})$, where λ is Lebesgue measure.

Proof. It suffices to prove (2.68) in $\mathbb{M}_p([0, T] \times [-\infty, \infty] \setminus \{0\})$ for any $T > 0$. To see this, observe that for $f \in \mathbb{C}_K^+([0,\infty) \times [-\infty, \infty] \setminus \{0\})$ with compact support in $[0, T] \times [-\infty, \infty] \setminus \{0\}$, the Laplace functional of a random measure at f is the same as the restriction of the random measure to $[0, T] \times [-\infty, \infty] \setminus \{0\}$ evaluated on the restriction of f to $[0, T] \times [-\infty, \infty] \setminus \{0\}$.

For convenience, we restrict our attention to proving convergence in $\mathbb{M}_+([0, 1] \times [-\infty, \infty] \setminus \{0\})$.

Suppose that $U_1, \cdots, U_n$ are i.i.d. $U(0, 1)$ random variables, which are independent of $\{X_j\}$, with order statistics

$$U_{1:n} \leq U_{2:n} \leq \cdots \leq U_{n:n}.$$

It is necessary to prove

$$\sum_{j=1}^{n} \delta_{(U_{j:n}, X_j/b_n)} \Rightarrow PRM(\lambda \times \nu) \tag{2.69}$$

in $\mathbb{M}_p([0,1] \times [-\infty, \infty] \setminus \{0\})$. First, from the independence of $\{U_j\}$ and $\{X_n\}$,

$$\sum_{j=1}^{n} \delta_{(U_{j:n}, X_j/b_n)} \stackrel{d}{=} \sum_{j=1}^{n} \delta_{(U_j, X_j/b_n)}$$

as random elements of $\mathbb{M}_p([0,1] \times [-\infty, \infty] \setminus \{0\})$. Thus we need to prove

$$\sum_{j=1}^{n} \delta_{(U_j, X_j/b_n)} \Rightarrow PRM(\lambda \times \nu)$$

in $\mathbb{M}_p([0,1] \times [-\infty, \infty] \setminus \{0\})$. However, because of independence,

$$n\mathbb{P}[(U_1, \frac{X_1}{b_n}) \in \cdot] = \lambda \times n\mathbb{P}[\frac{X_1}{b_n} \in \cdot] \Rightarrow \lambda \times v,$$

and therefore, from Lemma 2.5, (2.69) follows.

Let $d(\cdot, \cdot)$ be the vague metric on $\mathbb{M}_p([0,1] \times [-\infty, \infty] \setminus \{0\})$. From Slutsky's theorem, it is enough to prove that

$$d(\sum_{j=1}^{n} \delta_{(\frac{j}{n}, X_j/b_n)}, \sum_{j=1}^{n} \delta_{(U_{j:n}, X_j/b_n)}) \xrightarrow{\mathbb{P}} 0 \tag{2.70}$$

as $n \to \infty$. From the definition of the vague metric, it is enough to prove for $h \in \mathbb{C}_K^+([0,1] \times [-\infty, \infty] \setminus \{0\})$ that

$$|\sum_{j=1}^{n} h(\frac{j}{n}, X_j/b_n) - \sum_{j=1}^{n} h(U_{j:n}, X_j/b_n)| \xrightarrow{\mathbb{P}} 0 \tag{2.71}$$

in $\mathbb{R}$. Suppose the compact support of h is contained in $[0,1] \times \{x : |x| > \theta\}$ for some $\theta > 0$. We have

$$\sum_{j=1}^{n} |h(\frac{j}{n}, X_j/b_n) - h(U_{j:n}, X_j/b_n)| 1_{[|X_j|/b_n > \theta]}$$

$$\leq \omega_h(\sup_{j \leq n} |\frac{j}{n} - U_{j:n}|) \sum_{j=1}^{n} 1_{[|X_j|/b_n > \theta]},$$

where, as usual, $\omega_h(\eta)$ is the modulus of continuity of the uniformly continuous function h :

$$\omega_h(\eta) = \sup_{|x-y| \leq \eta} |h(x) - h(y)|,$$

and

$$\sum_{j=1}^{n} 1_{[|X_j|/b_n > \theta]} = \sum_{j=1}^{n} \delta_{X_j/b_n}(\{x > \theta\})$$

is stochastically bounded. So it is enough to prove that

$$\sup_{j\le n} |\frac{j}{n} - U_{j:n}| \xrightarrow{\mathbb{P}} 0.$$

However, from the Glivenko–Cantelli theorem

$$\sup_{0\le x\le 1} |\frac{1}{n}\sum_{j=1}^{n} 1_{[U_j\le x]} - x| \to 0 \text{ a.s.}$$

and hence, we obtain this lemma. □

Assume that $\{X_n\}_{n\ge1}$ is an sequence of i.i.d. random variables, which are regular varying with index $\alpha \in (0,2)$, and let

$$p = \lim_{x\to\infty} \frac{\mathbb{P}(X_1 > x)}{\mathbb{P}(|X_1| > x)}, \qquad q = 1-p = \lim_{x\to\infty} \frac{\mathbb{P}(X_1 < -x)}{\mathbb{P}(|X_1| > x)}.$$

The following theorem holds.

Theorem 2.6. *Suppose that*

$$\lim_{\varepsilon\downarrow 0} \limsup_{n\to\infty} \frac{n}{a_n^2}\mathbb{E}(X_1^2 1_{|X_1|\le a_n\varepsilon}) = 0, \tag{2.72}$$

where $\{a_n\}_{n\ge1}$ is a sequence of positive real numbers such that

$$n\mathbb{P}(|X_1| > a_n) \to 1.$$

Define the partial sum stochastic process

$$Z_n(t) = \frac{1}{a_n}\sum_{k=1}^{[nt]}(X_k - \mathbb{E}(X_k 1_{[|X_k|\le a_n]})), \quad t\ge 0.$$

Then

$$Z_n \Rightarrow Z_\alpha$$

in $\mathbb{D}[0,1]$ endowed with the J_1 topology, where $(Z_\alpha(t))_{t\in[0,1]}$ is an $\alpha-$stable Lévy process with Lévy triple $(b,0,\nu)$ given by

$$\nu(dx) = (p1_{(0,\infty)}(x) + q1_{(-\infty,0)}(x))\alpha|x|^{-\alpha-1}dx,$$

$$b = \int_{\{|x|\le 1\}} x\nu(dx).$$

Proof. From Lemma 2.6, we know that

$$\sum_{k=1}^{\infty}\delta_{(\frac{k}{n},\frac{X_k}{a_n})} \Rightarrow \sum_{k=1}^{\infty}\delta_{(t_k,j_k)} = PRM(\lambda\times\nu)$$

in $\mathbb{M}_p([0,\infty)\times[-\infty,\infty]\setminus\{0\})$. For convenience, we assume that 1 is not a jump of the function

$$\tau(t) = \nu\{x : |x| > t\}.$$

Here and in the rest of the proof, we always assume ε is chosen so that ε is not a jump point of $\tau(\cdot)$.

Define the restriction map

$$\mathbb{M}_p([0,\infty)\times[-\infty,\infty]\setminus\{0\}) \longmapsto \mathbb{M}_p([0,\infty)\times\{x:|x|>\varepsilon\})$$

by

$$m \longmapsto m|_{[0,\infty)\times\{x:|x|>\varepsilon\}}.$$

It is almost surely continuous with respect to the distribution of $PRM(\lambda\times\nu)$.

We claim that the *summation functional*

$$\psi^{(\varepsilon)}:\mathbb{M}_p([0,\infty)\times\{x:|x|>\varepsilon\}) \longmapsto \mathbb{D}([0,T],\mathbb{R})$$

defined by

$$\psi^{(\varepsilon)}(\sum_{k=1}^{\infty}\delta_{(\tau_k,J_k)})(t)\to\sum_{\tau_k\le t}J_k,$$

is continuous respect to the distribution of $PRM(\lambda\times\nu)$ almost surely.

Fix some $T>0$. We will show that if $m_1,m_2\in\mathbb{M}_p([0,\infty)\times\{x:|x|>\varepsilon\})$ are closed, then $\psi^{(\varepsilon)}(m_1)$ and $\psi^{(\varepsilon)}(m_2)$ are closed as functions in $\mathbb{D}[0,T]$. Define the subset of $\mathbb{M}_p([0,\infty)\times\{x:|x|>\varepsilon\})$

$$\begin{aligned}\Lambda=\{\,m\in\mathbb{M}_p([0,\infty)\times\{x:|x|>\varepsilon\}):\,&m([0,\infty)\times\{\pm\varepsilon\})=0,\\ &m([0,\infty)\times\{\pm\infty\})=0,\\ &m(\{0\}\times\{x:|x|>\varepsilon\})=m(\{T\}\times\{x:|x|>\varepsilon\})=0,\\ &m\{[0,T]\times\{x:|x|>\varepsilon\}\}<\infty,\\ &\text{and no vertical line contains two points of } m(([0,T]\times\{x:|x|>\varepsilon\})\cap\cdot)\}.\end{aligned}$$

Denote $PRM(\lambda\times\nu)$ by $\mathbb{N}$. We analyze Λ as an intersection of several sets and show that each of the intersecting sets has probability 1. First, we have

$$\mathbb{E}(\mathbb{N}([0,T]\times\{x:|x|>\varepsilon\}))=T\nu(\{x:|x|>\varepsilon\})<\infty,$$

so

$$\mathbb{P}[\mathbb{N}([0,T]\times\{x:|x|>\varepsilon\})<\infty]=1.$$

Second, we have

$$\lambda\times\nu(\{0\}\times\{x:|x|>\varepsilon\})=\lambda(\{0\})\cdot\nu(\{x:|x|>\varepsilon\})=0,$$

so

$$\mathbb{E}(\mathbb{N}(\{0\}\times\{x:|x|>\varepsilon\}))=0,$$

and therefore

$$\mathbb{P}[\mathbb{N}(\{0\}\times\{x:|x|>\varepsilon\})=0]=1.$$

Similarly,

$$\mathbb{P}[\mathbb{N}(\{T\} \times \{x : |x| > \varepsilon\}) = 0] = 1.$$

Next, we show

$$\mathbb{P}\{\text{no vertical line contains two points of } \mathbb{N}(([0,T] \times \{x : |x| > \varepsilon\}) \cap \cdot)\} = 1. \quad (2.73)$$

Pick any $M > 0$. We can represent

$$\mathbb{N}(([0,T] \times \{x : |x| > \varepsilon\}) \cap \cdot)\} = \sum_{i=1}^{\xi} \delta_{(U_i,V_i)}(\cdot),$$

where ξ is a Poisson random variable with parameter $T\nu(\{x : |x| > \varepsilon\})$, $\{U_i, i \geq 1\}$ are i.i.d. uniformly distributed on $(0,T)$, and $\{V_i, i \geq 1\}$ are i.i.d. with distribution $\nu(\{x : |x| > \varepsilon\} \cap \cdot)/\nu(\{x : |x| > \varepsilon\})$. Then

$$\begin{aligned}
&\mathbb{P}\{\text{some vertical line contains two points of } \mathbb{N}(([0,M] \times \{x : |x| > \varepsilon\}) \cap \cdot)\} \\
&= \mathbb{P}\left\{ \bigcup_{1\leq i<j\leq \xi} [U_i = U_j] \right\} \\
&= \sum_{n=0}^{\infty} \mathbb{P}\left\{ \bigcup_{1\leq i<j\leq n} [U_i = U_j] \right\} \mathbb{P}[\xi = n] \\
&\leq \sum_{n=0}^{\infty} \binom{n}{2} \mathbb{P}[U_1 = U_2]\mathbb{P}[\xi = n] \\
&= \sum_{n=0}^{\infty} \binom{n}{2} \cdot 0 \cdot \mathbb{P}[\xi = n] = 0.
\end{aligned}$$

This gives (2.73). Thus $\mathbb{P}[\mathbb{N} \in \Lambda] = 1$.

For continuity of $\psi^{(\varepsilon)}$, it is enough that $\psi^{(\varepsilon)}(m_n) \to \psi^{(\varepsilon)}(m)$ in $\mathbb{D}[0,1]$ according to the J_1 topology if $m_n \xrightarrow{v} m$ in $\mathbb{M}_p$ for some $m \in \Lambda$.

Since $m \in \Lambda$, there exists a nonnegative integer $k = k(m)$ such that

$$m([0,T] \times \{x : |x| > \varepsilon\}) = k < \infty.$$

By the assumption, η does not have any atoms on the horizontal lines at u or $-u$. Thus there exists a positive integer n_0 such that for all $n \geq n_0$,

$$m_n([0,T] \times \{x : |x| > \varepsilon\}) = k < \infty.$$

If $k = 0$, there is nothing to prove, so we assume $k \geq 1$. Let (t_i, x_i), $(t_i^{(n)}, x_i^{(n)})$ be the atoms of m and m_n in $[0,T] \times \{x : |x| > \varepsilon\}$ respectively. Pick γ so small that

$$t_1 - \gamma > \gamma, \quad t_i + \gamma < t_{i+1} - \gamma, \quad i = 1, 2, \cdots, k-1, \quad t_k + \gamma < T - \gamma$$

and there exists $n_1 > n_0$, such that for $n > n_1$,

$$(t_i^{(n)}, x_i^{(n)}) \in (t_i - \gamma, t_i + \gamma) \times (x_i - \gamma, x_i + \gamma), \quad 1 \leq i \leq k.$$

Define continuous one-to-one mapping $\lambda_n : [0,T] \to [0,T]$ by

$$\lambda_n(0) = 0, \quad \lambda_n(T) = T,$$

$$\lambda_n(t_i^{(n)}) = t_i \quad i = 1, \cdots, k,$$

between these points, λ_n is defined by linear interpolation. We have

$$\begin{aligned}
\sup_{t\in[0,T]} |\psi^{(\varepsilon)}(m_n)(\lambda_n^{-1}(t)) - \psi^{(\varepsilon)}(m_n)(t)| &= \sup_{t\in[0,T]} |\sum_{t_k^{(n)} \le \lambda_n^{-1}(t)} x_k^{(n)} - \sum_{t_k \le t} x_k| \\
&= \sup_{t\in[0,T]} |\sum_{\lambda_n(t_k^{(n)}) \le t} x_k^{(n)} - \sum_{t_k \le t} x_k| \\
&= \sup_{t\in[0,T]} |\sum_{t_k \le t} x_k^{(n)} - \sum_{t_k \le t} x_k| \\
&\le \sum_{t_k \le T} |x_k^{(n)} - x_k| \le k\gamma.
\end{aligned}$$

On the other hand,

$$\sup_{s\in[0,T]} |\lambda_n(s) - s| = \bigvee_{i=1}^{k} \bigvee_{s\in[t_i^{(n)}, t_{i+1}^{(n)}]} |\lambda_n(s) - s|.$$

On the first interval,

$$\begin{aligned}
\sup_{s\in[0,t_1^{(n)}]} |\lambda_n(s) - s| &= \sup_{s\in[0,t_1^{(n)}]} |\frac{t_1}{t_1^{(n)}} s - s| \\
&\le |\frac{t_1}{t_1^{(n)}} - 1| t_1^{(n)} = |t_1 - t_1^{(n)}| \le \gamma.
\end{aligned}$$

On $[t_i^{(n)}, t_{i+1}^{(n)}]$,

$$\lambda_n(s) = \frac{t_{i+1} - t_i}{t_{i+1}^{(n)} - t_i^{(n)}} s + t_i - \frac{t_{i+1} - t_i}{t_{i+1}^{(n)} - t_i^{(n)}} \cdot t_i^{(n)}.$$

Thus

$$\begin{aligned}
&\sup_{s\in[t_i^{(n)}, t_{i+1}^{(n)}]} |\lambda_n(s) - s| \\
&= \sup_{s\in[t_i^{(n)}, t_{i+1}^{(n)}]} |\frac{t_{i+1} - t_i}{t_{i+1}^{(n)} - t_i^{(n)}} s + t_i - \frac{t_{i+1} - t_i}{t_{i+1}^{(n)} - t_i^{(n)}} \cdot t_i^{(n)} - s| \\
&= \sup_{s\in[0, t_{i+1}^{(n)} - t_i^{(n)}]} |\frac{t_{i+1} - t_i}{t_{i+1}^{(n)} - t_i^{(n)}} \cdot (s + t_i^{(n)}) + t_i - \frac{t_{i+1} - t_i}{t_{i+1}^{(n)} - t_i^{(n)}} \cdot t_i^{(n)} - (s + t_i^{(n)})| \\
&= \sup_{s\in[0, t_{i+1}^{(n)} - t_i^{(n)}]} |(\frac{t_{i+1} - t_i}{t_{i+1}^{(n)} - t_i^{(n)}} - 1)s + t_i - t_i^{(n)}| \\
&\le |(\frac{t_{i+1} - t_i}{t_{i+1}^{(n)} - t_i^{(n)}} - 1)(t_{i+1}^{(n)} - t_i^{(n)})| + |t_i - t_i^{(n)}|
\end{aligned}$$

$$\leq |t_{i+1} - t_i| + |t_{i+1}^{(n)} - t_i^{(n)}| + \gamma \leq 3\gamma.$$

So

$$\sup_{s\in[0,T]} |\lambda_n(s) - s| \leq 3\gamma.$$

Thus $\psi^{(\varepsilon)}(m_n) \to \psi^{(\varepsilon)}(m)$ in $\mathbb{D}[0,1]$ according to the J_1 topology. Hence, $\psi^{(\varepsilon)}$ is continuous almost surely respect to the distribution of $\mathbb{N}$. Then, we have

$$\sum_{k=1}^{n} 1_{[|X_k|>a_n\varepsilon]}\epsilon_{(\frac{k}{n},\frac{X_k}{a_n})} \Rightarrow \sum_{k=1}^{\infty} 1_{[||j_k||>\varepsilon]}\epsilon_{(t_k,j_k)}$$

in $\mathbb{M}_p([0,\infty)\times\{x:|x|>\varepsilon\})$, and

$$\sum_{k=1}^{[n\cdot]} \frac{X_k}{a_n} 1_{[|X_k|>a_n\varepsilon]} \Rightarrow \sum_{t_k\leq(\cdot)} j_k 1_{[|j_k|>\varepsilon]}$$

in $\mathbb{D}([0,T])$. Similarly,

$$\sum_{k=1}^{[n\cdot]} \frac{X_k}{a_n} 1_{[\varepsilon<|\frac{X_k}{a_n}|\leq 1]} \Rightarrow \sum_{t_k\leq(\cdot)} j_k 1_{[\varepsilon<|j_k|\leq 1]}.$$

Then, taking expectations, we have

$$\frac{[n\cdot]}{a_n}\mathbb{E}(X_1 1_{[\varepsilon<|\frac{X_1}{a_n}|\leq 1]}) \Rightarrow (\cdot)\int_{\{x:\varepsilon<|x|\leq 1\}} x\nu(dx)$$

in $\mathbb{D}([0,T])$. To justify this, observe first for any $t>0$ that

$$\begin{aligned}\frac{[nt]}{a_n}\mathbb{E}(X_1 1_{[\varepsilon<|\frac{X_1}{a_n}|\leq 1]}) &= \frac{[nt]}{n}\int_{\{x:|x|\in(\varepsilon,1]\}} xn\mathbb{P}[\frac{X_1}{a_n}\in dx] \\ &\to t\int_{\{x:|x|\in(\varepsilon,1]\}} x\nu(dx)\end{aligned}$$

since $n\mathbb{P}[\frac{X_1}{a_n}\in\cdot] \stackrel{\nu}{\to} \nu(\cdot)$ and ε and 1 are not jumps of $\tau(\cdot)$. Convergence is locally uniform in t and hence convergence takes place in $\mathbb{D}([0,T])$. Thus

$$\begin{aligned}Z_n^{(\varepsilon)}(\cdot) &:= \sum_{k=1}^{[n\cdot]} \frac{X_k}{a_n} 1_{[|X_k|>a_n\varepsilon]} - [n\cdot]\mathbb{E}(\frac{X_1}{a_n} 1_{[\varepsilon<|\frac{X_1}{a_n}|\leq 1]}) \\ &\Rightarrow \sum_{t_k\leq(\cdot)} j_k 1_{[|j_k|>\varepsilon]} - (\cdot)\int_{\{x:|x|\in(\varepsilon,1]\}} x\nu(dx) =: Z_\alpha^{(\varepsilon)}(\cdot).\end{aligned}$$

From the Itô representation of Lévy process, for almost all ω, as $\varepsilon\downarrow 0$,

$$Z_\alpha^{(\varepsilon)}(\cdot) \to Z_\alpha(\cdot)$$

locally uniformly in t. Hence, it suffices to show that

$$\lim_{\varepsilon\downarrow 0}\limsup_{n\to\infty}\mathbb{P}\left[\sup_{0\leq t\leq T}|Z_n^{(\varepsilon)}(t) - Z_n(t)| > \delta\right] = 0 \tag{2.74}$$

for any $\delta > 0$. Recalling the definitions of $Z_n^{(\varepsilon)}$, we have

$$|Z_n^{(\varepsilon)}(t) - Z_n(t)| = \left|\sum_{k=1}^{[nt]} \frac{X_k}{a_n} 1_{[|X_k| \le a_n \varepsilon]} - [nt]\mathbb{E}(\frac{X_1}{a_n} 1_{[|X_1| \le a_n \varepsilon]})\right|$$

$$= \left|\sum_{k=1}^{[nt]} \left(\frac{X_k}{a_n} 1_{[|X_k| \le a_n \varepsilon]} - \mathbb{E}(\frac{X_k}{a_n} 1_{[|X_{n,k}| \le a_n \varepsilon]})\right)\right|,$$

so by Kolmogorov's inequality,

$$\mathbb{P}\left[\sup_{0 \le t \le T} |Z_n^{(\varepsilon)}(t) - Z_n(t)| > \delta\right]$$

$$= \mathbb{P}[\sup_{0 \le t \le T} \left|\sum_{k=1}^{[nt]} \left(\frac{X_k}{a_n} 1_{[|X_k| \le a_n \varepsilon]} - \mathbb{E}(\frac{X_k}{a_n} 1_{[|X_k| \le a_n \varepsilon]})\right)\right| > \delta]$$

$$= \mathbb{P}[\max_{0 \le j \le nT} \left|\sum_{k=1}^{j} \left(\frac{X_k}{a_n} 1_{[|X_k| \le \varepsilon]} - \mathbb{E}(\frac{X_k}{a_n} 1_{[|X_k| \le a_n \varepsilon]})\right)\right| > \delta]$$

$$\le \frac{1}{\delta^2} Var\left(\sum_{k=1}^{[nT]} \frac{X_k}{a_n} 1_{[|X_k| \le a_n \varepsilon]}\right)$$

$$= \frac{[nT]}{\delta^2} Var\left(\frac{X_1}{a_n} 1_{[|X_1| \le a_n \varepsilon]}\right)$$

$$\le \frac{[nT]}{\delta^2} \mathbb{E}\left(\left(\frac{X_1}{a_n}\right)^2 1_{[|X_1| \le a_n \varepsilon]}\right)$$

$$\to \int_{|x| \le \varepsilon} x^2 \nu(dx)$$

as $n \to \infty$. Letting $\varepsilon \to 0$, we obtain (2.74). The theorem is proved. □

2.3.2 *Functional limit theorem for dependent heavy-tailed sequence*

In the previous subsection, we discuss the functional limit theorem for independent heavy-tailed sequence by means of point process convergence method. In this subsection, we discuss the functional limit theorem of dependent heavy-tailed sequence with regular variation with index $\alpha \in (0, 2)$.

Similar to last subsection, we investigate the asymptotic distributional behavior of processes

$$Z_n(t) = \frac{1}{a_n}(\sum_{i=1}^{[nt]} X_i - [nt]b_n), \quad t \in [0, 1],$$

valued in $\mathbb{D}[0, 1]$, where $\{X_i\}_{i \ge 1}$ is a stationary sequence of random variables, $\{a_n\}_{n \ge 1}$ is a sequence of positive real numbers such that

$$n\mathbb{P}(|X_1| > a_n) \to 1$$

as $n \to \infty$ and $b_n = \mathbb{E}(X_1 1_{\{|X_1| \leq a_n\}})$.

In independent case, the weak convergence of Z_n holds in Skorohod's J_1 topology. For dependent case, the asymptotic behavior of Z_n is quite complex, due to the different dependent structures. Leadbetter and Rootzén (1988), Tyran-Kamińska (2010) studied that subject. Essentially, what they showed is that the functional limit theorem holds in Skorohod's J_1 topology iff certain point processes to a Poisson point process, which in turn is equivalent to a kind of nonclustering property for heavy tailed random variables. However, for many interesting models, convergence in the J_1 topology can not hold. For example, Avram and Taqqu (1992) showed that when $\{X_n\}_{n\in\mathbb{Z}}$ ($\mathbb{Z}$ is the integers set of $\mathbb{R}$) is linear process with regular varying innovations and nonnegative coefficients, $\{Z_n(\cdot)\}$ converges weakly in the M_1 topology instead. *Basrak, Krizmanić and Segers (2012)* showed that for a stationary, regularly varying sequence for which clusters of high-threshold excesses can be broken down into asymptotically independent blocks, $\{Z_n(\cdot)\}$ converges weakly to an $\alpha-$stable Lévy process in $\mathbb{D}[0,1]$ with M_1 topology. The method of proving convergence of point processes in M_1 topology is different from that in J_1 topology. In 2.3.1, we have already introduced the convergence of point processes in J_1 topology. For the completeness, we introduce Basrak's result in this subsection.

To express the dependence between $\{X_n\}_{n\in\mathbb{Z}}$, it is necessary to introduce the definition of jointly regularly varying. From now on,

Definition 2.4. For a strictly stationary sequence $(X_n)_{n\in\mathbb{Z}}$, if for any positive integer k, and any norm $||\cdot||$ on $\mathbb{R}^k$, there exists a random vector Θ on the unit sphere $\mathbb{S}^{k-1} = \{x \in \mathbb{R}^k : ||x|| = 1\}$, and $\alpha \in (0,\infty)$ such that for every $u \in (0,\infty)$,

$$\frac{\mathbb{P}(||\mathbf{X}|| > ux, \mathbf{X}/||\mathbf{X}|| \in \cdot)}{\mathbb{P}(||\mathbf{X}|| > x)} \xrightarrow{v} u^{-\alpha}\mathbb{P}(\Theta \in \cdot)$$

as $x \to \infty$, where $\mathbf{X} = (X_1, \cdots, X_k)$, we say that $\{X_n\}_{n\in\mathbb{Z}}$ is *jointly regularly varying* with index α.

Basrak and Segers (2009) provides a convenient characterization of jointly regularly variation: it is necessary and sufficient that there exists a strictly stationary process $\{Y_n\}_{n\in\mathbb{Z}}$ with $\mathbb{P}(|Y_1| > y) = y^{-\alpha}$ for $y > 1$ such that as $x \to \infty$,

$$\{(x^{-1}X_n)_{n\in\mathbb{Z}} | |X_1| > x\} \xrightarrow{fidi} \{Y_n\}_{n\in\mathbb{Z}},$$

where $\xrightarrow{fidi}$ denotes convergence of finite-dimensional distributions, $\{Y_n\}_{n\in\mathbb{Z}}$ is called the tail process of $\{X_n\}_{n\geq 1}$.

Theorem 2.7. *Let $\{X_n\}_{n\in\mathbb{Z}}$ be a strictly stationary sequence of random variables, jointly regular varying with index $\alpha \in (0,2)$, and with the tail process $\{Y_n\}_{n\geq 1}$ almost surely has no two values of the opposite sign. Furthermore, suppose that there exists a positive integer sequence $\{r_n\}_{n\geq 1}$ such that $r_n \to \infty$, $r_n/n \to 0$ as $n \to \infty$ and for every $u > 0$*

$$\lim_{m\to\infty} \limsup_{n\to\infty} \mathbb{P}(\max_{m\leq i\leq r_n} |X_i| > ua_n | |X_0| > ua_n) = 0, \tag{2.75}$$

and for every $f \in \mathbb{C}_K^+([0,1] \times [-\infty,\infty]\backslash\{0\})$,

$$\mathbb{E}[\exp\{-\sum_{i=1}^{n} f(\frac{i}{n}, \frac{X_i}{a_n})\}] - \prod_{k=1}^{k_n} \mathbb{E}[\exp\{-\sum_{i=1}^{r_n} f(\frac{k_n r_n}{n}, \frac{X_i}{a_n})\}] \to 0 \tag{2.76}$$

as $n \to \infty$, *where* $k_n = [n/r_n]$. *If* $1 \le \alpha < 2$, *also suppose that for all* $\delta > 0$,

$$\lim_{u \downarrow 0} \limsup_{n\to\infty} \mathbb{P}(\max_{1\le k\le n} |\sum_{i=1}^{k} (\frac{X_i}{a_n} 1_{\{|X_i| \le u a_n\}} - \mathbb{E}(\frac{X_i}{a_n} 1_{\{|X_i|\le u a_n\}}))| > \delta) = 0. \tag{2.77}$$

Then

$$Z_n \Rightarrow Z_\alpha$$

in $\mathbb{D}[0,1]$ *endowed with the* M_1 *topology, where* $(Z_\alpha(t))_{t\in[0,1]}$ *is an* $\alpha-$*stable Lévy process with Lévy triple* $(b, 0, \mu)$ *given by the limits*

$$\nu^{(u)} \xrightarrow{v} \nu, \qquad \int_{\{u<|x|\le 1\}} x\nu^{(u)}(dx) - \int_{\{u<|x|\le 1\}} x\mu(dx) \to b$$

as $u \downarrow 0$, with

$$\nu^{(u)}(x,\infty) = u^{-\alpha} \mathbb{P}(u \sum_{i\ge 0} Y_i 1_{\{|Y_i|>1\}} > x, \sup_{i\le -1} |Y_i| \le 1),$$

$$\nu^{(u)}(-\infty,-x) = u^{-\alpha} \mathbb{P}(u \sum_{i\ge 1} Y_i 1_{\{|Y_i|>1\}} < -x, \sup_{i\le -1} |Y_i| \le 1),$$

$$\mu(dx) = (p 1_{(0,\infty)}(x) + q 1_{(-\infty,0)}(x))\alpha|x|^{-\alpha-1}dx,$$

$$p = \lim_{x\to\infty} \frac{\mathbb{P}(X_1 > x)}{\mathbb{P}(|X_1| > x)}, \qquad q = 1 - p = \lim_{x\to\infty} \frac{\mathbb{P}(X_1 < -x)}{\mathbb{P}(|X_1| > x)}.$$

Proof. Define the point process

$$\mathbb{N}_n = \sum_{i=1}^{n} \delta_{(\frac{i}{n}, \frac{X_i}{a_n})}, \quad n \ge 1.$$

We first show that

$$\mathbb{N}_n|_{[0,1]\times[-\infty,-u)\cup(u,\infty]} \xrightarrow{d} \mathbb{N}^{(u)} := \sum_{i\ge 1}\sum_{j\ge 1} \delta_{(T_i^{(u)}, uZ_{ij})}|_{[0,1]\times[-\infty,-u)\cup(u,\infty]} \tag{2.78}$$

in $\mathbb{M}_p([0,1] \times [-\infty,-u) \cup (u,\infty])$, where

(1) $\sum_i \delta_{T_i^{(u)}}$ is a homogeneous Poisson process on $[0,1]$ with intensity θu^α;

(2) $(\sum_j \delta_{Z_{ij}})_i$ is an i.i.d. sequence of point processes in $\mathbb{R}$, independent of $\sum_i \delta_{T_i^{(u)}}$, and with common distribution equal to $(\sum_{n\ge 1} \delta_{Y_n} | \sup_{i\le -1} |Y_i| \le 1)$.

Let $\{X_{k,j}\}_{j\ge 1}$ with $k \ge 1$, be independent copies of $\{X_j\}_{j\ge 1}$, and define

$$\hat{\mathbb{N}}_n = \sum_{k=1}^{k_n} \hat{\mathbb{N}}_{n,k} \text{ with } \hat{\mathbb{N}}_{n,k} = \sum_{j=1}^{r_n} \delta_{(\frac{kr_n}{n}, \frac{X_{k,j}}{a_n})}.$$

By (2.76), the weak limits of $\mathbb{N}_n$ and $\hat{\mathbb{N}}_n$ must coincide. Similarly, it is enough to show that the Laplace functionals of $\hat{\mathbb{N}}_n$ converge to those of $\mathbb{N}^{(u)}$. Take $f \in \mathbb{C}_K^+([0,1] \times [-\infty,-u) \cup (u,\infty])$. We extend f to $[0,1] \times [-\infty,\infty]/\{0\}$ by setting $f(t,x) = 0$ whenever $|x| \leq u$, hence there exists $M > 0$, such that $0 \leq f(t,x) \leq M1_{[-u,u]^c}(x)$. Then

$$\begin{aligned}1 \geq \mathbb{E}[\exp(-\hat{\mathbb{N}}_{n,k}f)] &\geq \mathbb{E}[\exp(-M\sum_{i=1}^{r_n} 1_{\{|X_i|>a_n u\}})] \\ &\geq 1 - Mr_n\mathbb{P}(|X_0| > a_n u) = 1 - O(k_n^{-1})\end{aligned}$$

as $n \to \infty$. By $0 \leq -\log z - (1-z) \leq (1-z)^2/z$ for $z \in [0,1]$, it follows that

$$-\log \mathbb{E}[\exp(-\hat{\mathbb{N}}_n f)] = -\sum_{k=1}^{k_n} \log \mathbb{E}[\exp(-\hat{\mathbb{N}}_{n,k}f)] = \sum_{k=1}^{k_n}(1-\mathbb{E}[\exp(-\hat{\mathbb{N}}_{n,k}f)]) + O(k_n^{-1})$$

as $n \to \infty$. Furthermore, by Proposition 4.2 in Basrak and Segers (2009),

$$\begin{aligned}\theta := \mathbb{P}(\sup_{i\geq 1}|Y_i| \leq 1) &= \lim_{n\to\infty} \mathbb{P}(\max_{1\leq i\leq r_n} |X_i| \leq a_n u \,|\, |X_0| > a_n u) \\ &= \lim_{n\to\infty} \frac{\mathbb{P}(\max_{1\leq i\leq r_n} |X_i| \leq a_n u)}{r_n\mathbb{P}(|X_0| > a_n u)},\end{aligned}$$

and

$$\lim_{n\to\infty} \mathbb{E}[\sum_{i=1}^{r_n} 1_{(a_n u,\infty)}(X_i) \,|\, \max_{1\leq i\leq r_n} |X_i| > a_n u] = \frac{1}{\theta}. \tag{2.79}$$

Thus

$$\begin{aligned}&\sum_{k=1}^{k_n}(1 - \mathbb{E}[\exp(-\hat{\mathbb{N}}_{n,k}f)]) \\ &= k_n\mathbb{P}(\max_{1\leq i\leq r_n} |X_i| > a_n u)\frac{1}{k_n}\sum_{k=1}^{k_n}\mathbb{E}[1 - \exp(-\sum_{j=1}^{r_n} f(\frac{kr_n}{n}, \frac{X_j}{a_n}))| \max_{1\leq i\leq r_n} |X_i| > a_n u] \\ &= \theta u^{-\alpha}\frac{1}{k_n}\sum_{k=1}^{k_n}\mathbb{E}[1 - \exp(-\sum_{j=1}^{r_n} f(\frac{kr_n}{n}, \frac{X_j}{a_n}))| \max_{1\leq i\leq r_n} |X_i| > a_n u] + o(1).\end{aligned}$$

Let T_n be uniformly distributed random variables on $\{kr_n/n : k = 1, 2, \cdots, k_n\}$ and independent of $\{X_n\}_{n\in\mathbb{Z}}$, we have

$$\sum_{k=1}^{k_n}(1-\mathbb{E}[\exp(-\hat{\mathbb{N}}_{n,k}f)]) = \theta u^{-\alpha}\mathbb{E}[1-\exp(-\sum_{j=1}^{r_n} f(T_n, \frac{X_j}{a_n}))| \max_{1\leq i\leq r_n} |X_i| > a_n u] + o(1).$$

Note that T_n converges in law to random variable T with a uniformly distribute on $(0,1)$, by (2.79),

$$(T_n, \sum_{i=1}^{r_n} \delta_{\frac{X_i}{a_n}} | \max_{1\leq i\leq r_n} |X_i| > a_n u) \xrightarrow{d} (T, \sum_{n=1}^{\infty} \delta_{uZ_n})$$

where $\sum_n \delta_{Z_n}$ is a point process on $[-\infty, \infty]/\{0\}$, independent of T and with distribution equal to $(\sum_{n\geq 1}\delta_{Y_n} | \sup_{i\leq -1}|Y_i|\leq 1)$. Thus

$$\lim_{n\to\infty}\theta u^{-\alpha}\frac{1}{k_n}\sum_{k=1}^{k_n}\mathbb{E}[1-\exp(-\sum_{j=1}^{r_n}f(\frac{kr_n}{n},\frac{X_j}{a_n}))|\max_{1\leq i\leq r_n}|X_i|>a_nu]$$

$$=\theta u^{-\alpha}\mathbb{E}[1-\exp(-\sum_{j\geq 1}f(T,uZ_j))]=\theta u^{-\alpha}\int_0^1\mathbb{E}[1-\exp(-\sum_{j\geq 1}f(t,uZ_j))]dt.$$

In the other hand,

$$\begin{aligned}\mathbb{E}[-\exp(\mathbb{N}^{(u)}f)] &= \mathbb{E}[\exp(-\sum_{i\geq 1}\sum_{j\geq 1}f(T_i^{(u)},uZ_{ij}))]\\ &= \mathbb{E}[\prod_{i\geq 1}\mathbb{E}[-\sum_{j\geq 1}f(T_i^{(u)},uZ_{ij})|T_i^{(u)}]]\\ &= \mathbb{E}[\exp(\sum_{i\geq 1}\log\mathbb{E}[\exp(-\sum_{j\geq 1}f(T_i^{(u)},uZ_j))]])].\end{aligned}$$

Since $\sum_i \delta_{T_i^{(u)}}$ is a homogeneous Poisson process on $[0,1]$ with intensity $\theta u^{-\alpha}$, the Laplace functionals of $\hat{\mathbb{N}}_n$ converge to those of $\mathbb{N}^{(u)}$. (2.78) is obtained.

Note the fact that

$$u\sum_{j\geq 1}Z_{ij}1_{\{|Z_{ij}|>1\}}$$

are i.i.d. and almost surely finite. Define

$$\tilde{\mathbb{N}}^{(u)}=\sum_{i\geq 1}\delta_{(T_i^{(u)},u\sum_j Z_{ij}1_{\{|Z_{ij}|>1\}})}.$$

Obviously, $\tilde{\mathbb{N}}^{(u)}$ is a Poisson random measure with mean measure $\theta u^{-\alpha}\lambda\times F^{(u)}$, where λ is the Lebesgue measure, and $F^{(u)}$ is the distribution of $u\sum_{j\geq 1}Z_{1j}1_{\{|Z_{1j}|>1\}}$. But for $0\leq s\leq t\leq 1$ and $x>0$, using the fact that the distribution of $\sum_{j\geq 1}\delta_{Z_{1j}}$ is equal to the one of $(\sum_{n\geq 1}\delta_{Y_n}|\sup_{i\leq -1}|Y_i|\leq 1)$, then

$$\begin{aligned}&\theta u^{-\alpha}\lambda\times F^{(u)}([s,t]\times(x,\infty))\\ &=\theta u^{-\alpha}(t-s)F^{(u)}((x,\infty))\\ &=\theta u^{-\alpha}(t-s)\mathbb{P}(u\sum_{j\geq 1}Z_{1j}1_{\{|Z_{1j}|>1\}}>x)\\ &=\theta u^{-\alpha}(t-s)\mathbb{P}(u\sum_{j\geq 1}Y_j1_{\{|Y_j|>1\}}>x|\sup_{i\leq -1}|Y_i|\leq 1)\\ &=\theta u^{-\alpha}(t-s)\frac{\mathbb{P}(u\sum_{j\geq 1}Y_j1_{\{|Y_j|>1\}}>x,\sup_{i\geq 1}|Y_i|\leq 1)}{\mathbb{P}(\sup_{i\leq -1}|Y_i|\leq 1)}\\ &=u^{-\alpha}(t-s)\mathbb{P}(u\sum_{j\geq 1}Y_j1_{\{|Y_j|>1\}}>x,\sup_{i\geq 1}|Y_i|\leq 1)\end{aligned}$$

$$= \lambda \times \nu^{(u)}([s,t] \times (x,\infty)).$$

The same can be done for $[s,t] \times (-\infty,-x)$, so $\theta u^{-\alpha}\lambda \times F^{(u)} = \lambda \times \nu^{(u)}$.

Consider the summation functional

$$\psi^{(u)} : \mathbb{M}_p([0,1] \times [-\infty,-u) \cup (u,\infty]) \to \mathbb{D}[0,1]$$

defined by

$$\psi^{(u)}(\sum_{i\geq 1} \delta_{(t_i,x_i)})(t) = \sum_{t_i \leq t} x_i 1_{\{u<x_i<\infty\}}.$$

As the proof in subsection 2.3.1, we need to prove that

$$\mathbb{P}(\mathrm{N}^{(u)} \in \Lambda) = 1, \tag{2.80}$$

and $\psi^{(u)}$ is continuous on the set Λ, when $\mathbb{D}[0,1]$ is endowed with M_1 topology, where $\Lambda = \Lambda_1 \bigcap \Lambda_2$,

$$\begin{aligned}\Lambda_1 = \{\eta \in \mathbb{M}_p([0,1] \times [-\infty,-u) \cup (u,\infty]) &: \eta(\{0,1\} \times [-\infty,-u) \cup (u,\infty]) \\ &= \eta([0,1] \times \{\pm\infty, \pm u\}) = 0\},\end{aligned}$$

$$\begin{aligned}\Lambda_2 = \{\eta \in \mathbb{M}_p([0,1] \times [-\infty,-u) \cup (u,\infty]) &: \eta(\{t\} \times (u,\infty]) \wedge \eta(\{t\} \times [-\infty,-u)\}) \\ &= 0 \text{ for all } t \in [0,1]\}.\end{aligned}$$

In fact, by the spectral decomposition $Y_i = |Y_0|\Theta_i$ into independent components Y_0 and Θ_i with $|Y_1|$ as Pareto random variable (c.f. Basrak and Segers (2009)), thus Y_i cannot have any atoms except at origin, and $\sum_{j\geq 1} \delta_{uY_j}(\{\pm u\}) = 0$, then $\sum_{j\geq 1} \delta_{uZ_{ij}}(\{\pm u\}) = 0$ as well. Together with $\mathbb{P}(\sum_{i\geq 1} \delta_{T_i^{(u)}}(\{0,1\}) = 0) = 1$, it implies $\mathbb{P}(\mathrm{N}^{(u)} \in \Lambda_1) = 1$.

Second, the assumption that with probability one the tail process has no two values of the opposite sign implies $\mathbb{P}(\mathrm{N}^{(u)} \in \Lambda_2) = 1$. Thus (2.80) holds.

For the continuity of $\psi^{(u)}$, it is enough to show that $\psi^{(u)}(\eta_n) \to \psi^{(u)}(\eta)$ in $\mathbb{D}[0,1]$ according to the M_1 topology if $\eta_n \xrightarrow{v} \eta$ in $\mathbb{M}_p$ for some $\eta \in \Lambda$.

Since $\eta \in \Lambda$, there exists a nonnegative integer $k = k(\eta)$ such that

$$\eta([0,1] \times [-\infty,-u) \cup (u,\infty]) = k < \infty.$$

By assumption, η does not have any atoms on the horizontal lines at u or $-u$. Thus, there exists a positive integer n_0 such that for all $n \geq n_0$,

$$\eta_n([0,1] \times [-\infty,-u) \cup (u,\infty]) = k < \infty.$$

If $k = 0$, there is nothing to prove, so we assume $k \geq 1$. Let (t_i, x_i), $(t_i^{(n)}, x_i^{(n)})$ be the atoms of η and η_n in $[0,1] \times [-\infty,-u) \cup (u,\infty]$ respectively. It is easy to see that, for every $\gamma > 0$, we can find a positive integer $n_\gamma > n_0$ such that for all $n \geq n_\gamma$,

$$|t_i^{(n)} - t_i| \leq \gamma, \quad |x_i^{(n)} - x_i| \leq \gamma \quad \text{for } i = 1,2,\cdots,k.$$

Let

$$0 < \tau_1 < \cdots < \tau_p < 1$$

be a sequence such that the sets $\{\tau_1, \cdots, \tau_p\}$ and $\{t_1, \cdots, t_k\}$ coincide. Since η can have several atoms with the same time coordinate, note that $p \le k$. Put $\tau_0 = 0$, $\tau_{p+1} = 1$, and set

$$0 < r < \frac{1}{2} \min_{0 \le i \le p} |\tau_{i+1} - \tau_i|.$$

For any $t \in [0,1] \setminus \{\tau_1, \cdots, \tau_p\}$, there exists $\epsilon \in (0, u)$ such that

$$\epsilon < r, \quad \epsilon \le \min_{1 \le i \le p} |t - \tau_i|.$$

Hence $n \ge n_\epsilon$ implies that $t_i^n \le t$ is equivalent to $t_i \le t$,

$$|\psi^{(u)}(\eta_n)(t) - \psi^{(u)}(\eta)(t)| = |\sum_{t_i^{(n)} \le t} x_i^{(n)} - \sum_{t_i \le t} x_i| \le \sum_{t_i \le t} \epsilon \le k\epsilon.$$

Put $v_i = \tau_i + r, \quad i \in \{1, 2, \cdots, p\}$.

For any $\epsilon < u \wedge r$, it is obvious that the function $\psi^{(u)}(\eta_n)$ $(n \ge n_\epsilon)$ and $\psi^{(u)}(\eta)$ are monotone on each of the intervals $[0, v_1], [v_1, v_2], \cdots, [v_p, 1]$. By the Proposition 1.1, $\psi^{(u)}(\eta_n) \xrightarrow{M_1} \psi^{(u)}(\eta)$. Thus $\psi^{(u)}$ is continuous on the set Λ.

Note that

$$\psi^{(u)}(\mathbb{N}_n|_{[0,1]\times[-\infty,-u)\cup(u,\infty]})(\cdot) = \sum_{i/n \le \cdot} \frac{X_i}{a_n} 1_{\{|X_i| > a_n u\}},$$

$$\psi^{(u)}(\mathbb{N}^{(u)})(\cdot) = \sum_{T_i^{(u)} \le \cdot} \sum_{j \ge 1} u Z_{ij} 1_{\{|Z_{ij}| > 1\}}$$

and

$$\psi^{(u)}(\mathbb{N}^{(u)}) = \psi^{(u)}(\tilde{\mathbb{N}}^{(u)}) \stackrel{d}{=} \psi^{(u)}(\bar{\mathbb{N}}^{(u)}),$$

where

$$\bar{\mathbb{N}}^{(u)} = \sum_{i \ge 1} \delta_{(T_i, K_i^{(u)})}$$

is a Poisson random measure with mean measure $\lambda \times \nu^{(u)}$. We obtain

$$\sum_{i=1}^{[n\cdot]} \frac{X_i}{a_n} 1_{\{|X_i| > a_n u\}} \Rightarrow \sum_{T_i \le \cdot} K_i^{(u)}$$

as $n \to \infty$ in $\mathbb{D}[0,1]$ under M_1 metric. As proved in Theorem 2.6,

$$[n\cdot]\mathbb{E}(\frac{X_1}{a_n} 1_{\{u < |\frac{X_1}{a_n}| \le 1\}}) \to (\cdot) \int_{\{x: u < |x| \le 1\}} x \mu(dx).$$

Since $(\cdot)\int_{\{x:u<|x|\le 1\}} x\mu(dx)$ is a continuous function, we obtain

$$\begin{aligned}
Z_n^{(u)}(\cdot) &= \sum_{i=1}^{[n\cdot]} \frac{X_i}{a_n} 1_{\{|X_i|>a_n u\}} - [n\cdot]\mathbb{E}(\frac{X_1}{a_n} 1_{\{u<|\frac{X_1}{a_n}|\le 1\}}) \\
\Rightarrow Z_\alpha^{(u)}(\cdot) &= \sum_{T_i\le \cdot} K_i^{(u)} - (\cdot)\int_{\{x:u<|x|\le 1\}} x\mu(dx), \\
&= \sum_{T_i\le \cdot} K_i^{(u)} - (\cdot)\int_{\{x:u<|x|\le 1\}} x\nu^{(u)}(dx) \\
&\quad +(\cdot)(\int_{\{x:u<|x|\le 1\}} x\nu^{(u)}(dx) - \int_{\{x:u<|x|\le 1\}} x\mu(dx)).
\end{aligned}$$

By the Lévy-Itô representation of Lévy process,

$$Z_\alpha^{(u)}(\cdot) \Rightarrow Z_\alpha(\cdot)$$

as $u \to 0$. Thus,

$$Z_n^{(u)}(\cdot) \Rightarrow Z_\alpha(\cdot)$$

as $u \to 0$. In fact

$$\begin{aligned}
&\lim_{u\to 0}\limsup_{n\to\infty} \mathbb{P}(\sup_{0\le t\le 1} |Z_n^{(u)}(t) - Z_n(t)| > \delta) \\
&= \lim_{u\to 0}\limsup_{n\to\infty} \mathbb{P}(\sup_{0\le t\le 1} |\sum_{i=1}^{[n\cdot]} (\frac{X_i}{a_n} 1_{\{|X_i|\le a_n u\}} - \mathbb{E}(\frac{X_i}{a_n} 1_{\{|\frac{X_i}{a_n}|\le u\}}))| > \delta) \\
&= \lim_{u\to 0}\limsup_{n\to\infty} \mathbb{P}(\max_{1\le k\le n} |\sum_{i=1}^{k} (\frac{X_i}{a_n} 1_{\{|X_i|\le a_n u\}} - \mathbb{E}(\frac{X_i}{a_n} 1_{\{|\frac{X_i}{a_n}|\le u\}}))| > \delta).
\end{aligned}$$

Note that for $\alpha \in (0,1)$, we have

$$\mathbb{P}(\max_{1\le k\le n} |\sum_{i=1}^{k} (\frac{X_i}{a_n} 1_{\{|X_i|\le a_n u\}} - \mathbb{E}(\frac{X_i}{a_n} 1_{\{|\frac{X_i}{a_n}|\le u\}}))| > \delta) \le \frac{2n}{a_n\delta}\mathbb{E}[|X_1| 1_{\{|X_1|\le u a_n\}}]$$

and for $\alpha \in [1,2)$,

$$\limsup_{n\to\infty} \frac{2n}{a_n\delta}\mathbb{E}[|X_1| 1_{\{|X_1|\le u a_n\}}] = \frac{\alpha}{1-\alpha} u^{1-\alpha}$$

by Karamata's theorem. Hence (2.77) holds. Thus the theorem is proved. □

2.4 Two examples of applications of point process method

The point process method is a powerful method in the heavy-tailed theory. To illustrate this, we will introduce some applications of point process method in the studying of time series with heavy tail.

From the previous section, we show that the asymptotic properties of regular varying sequence in the case of $\alpha \in [1, 2)$ are more complicated than in the case of $\alpha \in (0, 1)$. For the sake of simplicity, in this section, we will assume that $\{X_n\}_{n\geq 1}$ is a i.i.d. sequence of random variables, regular varying with index $\alpha \in (0, 1)$, and let

$$p = \lim_{x\to\infty} \frac{\mathbb{P}(X_1 > x)}{\mathbb{P}(|X_1| > x)}, \quad q = 1 - p = \lim_{x\to\infty} \frac{\mathbb{P}(X_1 < -x)}{\mathbb{P}(|X_1| > x)}.$$

Set $\{a_n\}_{n\geq 1}$ is a sequence of positive real numbers such that

$$n\mathbb{P}(|X_1| > a_n) \to 1.$$

In fact, $a_n = n^{1/\alpha}L(n)$, where $L(n)$ is a slowly varying function.

In what follows, three independent sequence $\{\Gamma_n\}_{n\geq 1}$, $\{U_n\}_{n\geq 1}$, $\{B_n\}_{n\geq 1}$ are defined on the same probability space. $\{\Gamma_n\}_{n\geq 1}$ is the arrival sequence of a unit rate Poisson process on $\mathbb{R}_+$. $\{U_n\}_{n\geq 1}$ is a i.i.d. sequence of $U(0, 1)$ variables, and $\{B_n\}_{n\geq 1}$ is i.i.d sequence satisfying

$$\mathbb{P}[B_1 = 1] = p, \quad \mathbb{P}[B_1 = -1] = q.$$

2.4.1 *The maximum of the periodogram for heavy-tailed sequence*

Recall that $\{X_n\}_{n\geq 1}$ is a i.i.d. sequence of random variables, regular varying with index $\alpha \in (0, 1)$, its *periodogram* is defined by

$$I_{n,X}(x) = |J_{n,X}(x)|^2 = \frac{1}{a_n} \sum_{j=1}^{n} X_j \exp(-\mathrm{i}2\pi xj)|^2 \quad x \in [0, \frac{1}{2}].$$

The periodogram is an estimate of the spectral density of a signal in practice. The statisticians are interested in the limit behavior of the sequence

$$M_{n,X} = \max_{x\in[0,\frac{1}{2}]} I_{n,X}(x) \text{ and } \widetilde{M}_{n,X} = \max_{j=1,2,\cdots,q} I_{n,X}(\omega_j),$$

where $2\omega_j = \frac{2j}{n}$, $j = 1, \cdots, q$, $q = q_n = \max\{j : 1 \leq j < \frac{n}{2}\}$.

Theorem 2.8. *(Mikosch, Resnick and Samorodnitsky (2000)) The limit relations*

$$M_{n,X} \Rightarrow Y_\alpha^2, \quad \widetilde{M}_{n,X} \Rightarrow Y_\alpha^2 \tag{2.81}$$

hold, where $Y_\alpha = \sum_{j=1}^{\infty} \Gamma_j^{-\frac{1}{\alpha}}$.

Proof. Note that

$$M_{n,X} \leq (\sum_{j=1}^{n} \frac{|X_j|}{a_n})^2, \tag{2.82}$$

and it is well known (c.f. *Feller (1971)*) that

$$\sum_{j=1}^{n} \frac{|X_j|}{a_n} \Rightarrow Y_\alpha.$$

Hence the sequence $\{M_{n,X}\}$ is stochastically bounded and it remains to show the lower bound in the limit for the maximum of the periodogram. It suffices to find

$$\liminf_{n\to\infty} \mathbb{P}[M_{n,X} > \gamma] \geq \mathbb{P}[Y_\alpha^2 > \gamma], \tag{2.83}$$

since then we would have

$$\begin{aligned}\mathbb{P}[Y_\alpha^2 > \gamma] &\leq \liminf_{n\to\infty} \mathbb{P}[M_{n,X} > \gamma] \leq \limsup_{n\to\infty} \mathbb{P}[M_{n,X} > \gamma] \\ &\leq \limsup_{n\to\infty} \mathbb{P}[(\sum_{j=1}^{n} \frac{|X_j|}{a_n})^2 > \gamma] = \mathbb{P}[Y_\alpha^2 > \gamma].\end{aligned}$$

For any $T > 0$ and $n \geq 2T$,

$$\begin{aligned}M_{n,X} &= \sup_{x\in[0,\frac{1}{2}]} |J_{n,X}(x)|^2 = \sup_{x\in[0,\frac{n}{2}]} |J_{n,X}(x/n)|^2 \\ &\geq \sup_{x\in[0,T]} |J_{n,X}(x/n)|^2.\end{aligned}$$

To show (2.83), we need to study the weak convergence of $J_{n,X}(x/n)$. Define the point processes

$$\mathbb{N}_n = \sum_{j=1}^{n} \delta_{(\frac{j}{n},\frac{X_j}{a_n})} \quad \text{and} \quad \mathbb{N} = \sum_{j=1}^{\infty} \delta_{(U_j, B_j \Gamma_j^{-\frac{1}{\alpha}})}.$$

By Lemma 2.6,

$$\mathbb{N}_n \xrightarrow{d} \mathbb{N}$$

in $\mathbb{M}_p([0,1]\times[-\infty,\infty]\setminus\{0\})$.

Pick $\eta > 0$, define

$$T_\eta : \mathbb{M}_p([0,1]\times[-\infty,\infty]\setminus\{0\}) \to \mathbb{C}[0,\infty)$$

by

$$T_\eta(\sum_{k\geq 1} \delta_{(t_k,x_k)})(x) = \sum_{k\geq 1} x_k 1_{\{|x_k|>\eta\}} \exp(-2\pi i x t_k).$$

If $y_n \to y$, $y_n, y \in [0,\infty)$, and $m_n \xrightarrow{v} m$, $m_n, m \in \mathbb{C}[0,\infty)$, we firstly prove

$$T_\eta(m_n)(y_n) \to T_\eta(m)(y). \tag{2.84}$$

To do this, denote

$$m_n = \sum_{k\geq 1} \delta_{(t_k^{(n)}, x_k^{(n)})}, \qquad m = \sum_{k\geq 1} \delta_{(t_k,x_k)}.$$

The set

$$K_\eta := [0,1]\times[-\infty,-\eta)] \cup [\eta,\infty]$$

is compact in $[0,1]\times[-\infty,\infty]\setminus\{0\}$ with $m(\partial K_\eta) = 0$. There exists n_0 such that for $n \geq n_0$,

$$m_n(K_\eta) = m(K_\eta) =: l,$$

and there is an enumeration of the points in K_η such that

$$((t_k^{(n)}, x_k^{(n)}), 1 \le k \le l) \to ((t_k, x_k), 1 \le k \le l),$$

and without loss of generality we may assume for given $b > 0$ that

$$\sup_{n \ge n_0} |y_n| \vee \max_{1 \le k \le l} |x_k^{(n)}| \le b.$$

Therefore,

$$\begin{aligned}
&|T_\eta(m_n)(y_n) - T_\eta(m)(y)| \\
&\le \sum_{k=1}^{l} |x_k^{(n)} \exp(-2\pi \mathrm{i} y_n t_k^{(n)}) - x_k \exp(-2\pi \mathrm{i} y t_k)| \\
&\le \sum_{k=1}^{l} |x_k^{(n)} \exp(-2\pi \mathrm{i} y_n t_k^{(n)}) - x_k \exp(-2\pi \mathrm{i} y_n t_k^{(n)}) \\
&\qquad + x_k \exp(-2\pi \mathrm{i} y_n t_k^{(n)}) - x_k \exp(-2\pi \mathrm{i} y t_k)| \\
&\le \sum_{k=1}^{l} |x_k^{(n)} - x_k| + \sum_{k=1}^{l} |x_k| |\exp(-2\pi \mathrm{i} y_n t_k^{(n)}) - \exp(-2\pi \mathrm{i} y t_k)| \\
&\to 0,
\end{aligned}$$

hence (2.84) holds. This means that the map T_η is continuous a.s. with respect to the distribution of $\mathbb{N}$.

Applying the functional T_η to $\mathbb{N}_n$ and $\mathbb{N}$, we have

$$J_{n,X}^{(\eta)}(x/n) := \sum_{j=1}^{n} \frac{X_j}{a_n} \exp(-2\pi \mathrm{i} x j/n) 1_{\{|X_j| \ge \eta a_n\}} \Rightarrow J_\infty^{(\eta)}(x) \tag{2.85}$$

in $\mathbb{C}[0, \infty)$, where

$$J_\infty^{(\eta)}(x) = \sum_{j=1}^{\infty} B_j \Gamma_j^{-\frac{1}{\alpha}} \exp(-2\pi \mathrm{i} x U_j) 1_{\{\Gamma_j^{-\frac{1}{\alpha}} > \eta\}}.$$

Let $\eta \to 0$,

$$J_\infty^{(\eta)}(x) \Rightarrow J_\infty(x) := \sum_{j=1}^{\infty} B_j \Gamma_j^{-\frac{1}{\alpha}} \exp(-2\pi \mathrm{i} x U_j).$$

We have

$$\begin{aligned}
&\lim_{\eta \to 0} \limsup_{n \to \infty} \mathbb{P}[\sup_{0 \le x \le 1} |J_{n,X}^{(\eta)}(x/n) - J_{n,X}(x)| > \delta] \\
&\le \lim_{\eta \to 0} \limsup_{n \to \infty} \mathbb{P}[\sup_{0 \le x \le 1} \sum_{j=1}^{n} \frac{X_j}{a_n} \exp(-2\pi \mathrm{i} x j/n) 1_{\{|X_j| < \eta a_n\}} > \delta] \\
&\le \lim_{\eta \to 0} \limsup_{n \to \infty} n \mathbb{E}[|\frac{X_1}{a_n}| 1_{\{|X_1| < \eta a_n\}}]/\delta = 0
\end{aligned}$$

by Karamata's theorem. Hence

$$J_{n,X}(x/n) \Rightarrow J_\infty(x) \tag{2.86}$$

holds in $\mathbb{C}[0,\infty)$.

Therefore,

$$\begin{aligned}\liminf_{n\to\infty} \mathbb{P}[M_{n,X} > \gamma] &\geq \liminf_{n\to\infty} \mathbb{P}[\sup_{x\in[0,T]} |J_{n,X}(x/n)|^2 > \gamma] \\ &= \mathbb{P}[\sup_{x\in[0,T]} |J_\infty(x)|^2 > \gamma].\end{aligned}$$

This is true for any T, so

$$\liminf_{n\to\infty} \mathbb{P}[M_{n,X} > \gamma] \geq \mathbb{P}[\sup_{x\in[0,\infty]} |J_\infty(x)|^2 > \gamma].$$

where

$$\begin{aligned}\sup_{x\in[0,\infty]} |J_\infty(x)|^2 &= \sup_{x\in[0,\infty]} |\sum_{j=1}^{\infty} B_j \Gamma_j^{-\frac{1}{\alpha}} \exp(-2\pi \mathrm{i} x U_j)|^2 \\ &\leq (\sum_{j=1}^{\infty} \Gamma_{ij}^{-\frac{1}{\alpha}})^2.\end{aligned}$$

In the other hand, define

$$\Upsilon = \{\omega \in \Omega : \sum_{j=1}^{\infty} \Gamma_j^{-\frac{1}{\alpha}}(\omega) < \infty \text{ and for every } \ m \geq 1 \text{ the numbers}$$
$$(U_1(\omega), \cdots, U_m(\omega)) \text{ are rationally independent}\}.$$

By the result of Weyl (1916), $\mathbb{P}[\Upsilon] = 1$, and

$$\{(\{xU_1(\omega)\}, \cdots, \{xU_m(\omega)\}), x \geq 0\}$$

(where $\{a\}$ denotes the fractional part of a) is dense in $[0,1]^m$.

Fix an element $\omega \in \Upsilon$ and for any $\varepsilon > 0$, there exists $N \geq 1$, such that

$$\sum_{j=N+1}^{\infty} \Gamma_j^{-\frac{1}{\alpha}} < \varepsilon.$$

There is a $x_0 \in [0,\infty)$ such that

$$\Re(B_j \exp(-2\pi \mathrm{i} x_0 U_j)) \geq 1 - \frac{\varepsilon}{N\Gamma_j^{-\frac{1}{\alpha}}}, \quad j = 1, \cdots, N.$$

Then

$$\begin{aligned}&\sup_{x\in[0,\infty]} |\sum_{j=1}^{\infty} B_j \Gamma_j^{-\frac{1}{\alpha}} \exp(-2\pi \mathrm{i} x U_j)| \\ \geq\ &\sup_{x\in[0,\infty]} |\sum_{j=1}^{N} B_j \Gamma_j^{-\frac{1}{\alpha}} \exp(-2\pi \mathrm{i} x U_j)| - \sum_{j=N+1}^{\infty} \Gamma_j^{-\frac{1}{\alpha}}\end{aligned}$$

$$\geq |\sum_{j=1}^{N} B_j \Gamma_j^{-\frac{1}{\alpha}} \exp(-2\pi i x_0 U_j)| - \varepsilon \geq \Re(\sum_{j=1}^{N} B_j \Gamma_j^{-\frac{1}{\alpha}} \exp(-2\pi i x_0 U_j)) - \varepsilon$$

$$\geq \sum_{j=1}^{N} (1 - \frac{\varepsilon}{N\Gamma_j^{-\frac{1}{\alpha}}}) \Gamma_j^{-\frac{1}{\alpha}} - \varepsilon = \sum_{j=1}^{N} \Gamma_j^{-\frac{1}{\alpha}} - \varepsilon.$$

Letting first $N \to \infty$ and then $\varepsilon \to 0$, we obtain

$$\sup_{x \in [0,\infty]} |J_\infty(x)|^2 \geq (\sum_{j=1}^{\infty} \Gamma_j^{-\frac{1}{\alpha}})^2.$$

Hence

$$\sup_{x \in [0,\infty]} |J_\infty(x)|^2 = (\sum_{j=1}^{\infty} \Gamma_j^{-\frac{1}{\alpha}})^2. \tag{2.87}$$

It remains to prove $\widetilde{M}_{n,X} \Rightarrow Y_\alpha^2$. In view of $M_{n,X} \Rightarrow Y_\alpha^2$, it suffices to show that for all $\gamma > 0$,

$$\liminf_{n \to \infty} \mathbb{P}[\widetilde{M}_{n,X} > \gamma] \geq \mathbb{P}[Y_\alpha^2 > \gamma].$$

In fact, for any fixed integer K and sufficiently large n,

$$\mathbb{P}[\max_{1 \leq j \leq q} |J_{n,X}(j/n)|^2 > \gamma] \geq \mathbb{P}[\max_{1 \leq j \leq K} |J_{n,X}(j/n)|^2 > \gamma],$$

and from (2.86),

$$(J_{n,X}(j/n), 1 \leq j \leq K) \xrightarrow{d} (J_\infty(j), 1 \leq j \leq K)$$

in $\mathbb{R}^K$, hence

$$\max_{1 \leq j \leq K} |J_{n,X}(j/n)|^2 \xrightarrow{d} \max_{1 \leq j \leq K} |J_\infty(j)|^2$$

and so

$$\liminf_{n \to \infty} \mathbb{P}[\max_{1 \leq j \leq q} |J_{n,X}(j/n)|^2 > \gamma] \geq \mathbb{P}[\sup_{j \geq 1} |J_\infty(j)|^2 > \gamma].$$

By (2.87), we finish the proof. □

2.4.2 *The weak convergence to stable integral*

Assume that for $0 < M < \infty$, $\beta \in [-M, M]$, $s \in [0,1]$, $f_n(\beta, s)$, $f(\beta, s)$ are continuous in β, $f(\beta, s)$ is differentiable in β and s, and for all $\beta \in [-M, M]$ and $s \in [0,1]$, there exists a constant C (depending on M), such that,

$$|f_n(\beta, s)| \leq C, \quad |f(\beta, s)| \leq C \quad \left|\frac{\partial f(\beta, s)}{\partial s}\right| \leq C, \quad \left|\frac{\partial f(\beta, s)}{\partial \beta}\right| \leq C$$

and

$$\sup_{\beta \in [-M,M], s \in [0,1]} |f_n(\beta, s) - f(\beta, s)| = O\left(\frac{1}{n}\right).$$

Recall that $\{X_n\}_{n\geq 1}$ is a sequence of i.i.d. random variables, regular varying with index $\alpha \in (0,1)$, we have

Theorem 2.9. *Let* $U_n(s) = a_n^{-1}\sum_{i=1}^{[ns]} X_i$, *and assume*

$$U_n \Rightarrow Z_\alpha$$

on $\mathbb{D}[0,1]$. *Then*

$$\sum_{i=1}^{n} f_n\left(\beta, \frac{i}{n}\right)\frac{X_i}{a_n} \Rightarrow \int_0^1 f(\beta, s)dZ_\alpha(s)$$

holds on $\mathbb{C}(\mathbb{R})$.

Proof. We first prove under local uniform metric,

$$\sum_{i=1}^{n} f_n\left(\beta, \frac{i}{n}\right)\frac{X_i}{a_n} - \sum_{i=1}^{n} f\left(\beta, \frac{i}{n}\right)\frac{X_i}{a_n} \xrightarrow{\mathbb{P}} 0$$

on $\mathbb{C}(\mathbb{R})$. In fact

$$\begin{aligned}
&\left|\sum_{i=1}^{n} f_n\left(\beta, \frac{i}{n}\right)\frac{X_i}{a_n} - \sum_{i=1}^{n} f\left(\beta, \frac{i}{n}\right)\frac{X_i}{a_n}\right| \\
&\leq \sum_{i=1}^{n}\left| f_n\left(\beta, \frac{i}{n}\right) - f\left(\beta, \frac{i}{n}\right)\right|\left|U_n\left(\frac{i}{n}\right) - U_n\left(\frac{i-1}{n}\right)\right|.
\end{aligned}$$

Let $\delta = \frac{\alpha}{m}$, $m > 1 + \frac{1}{\alpha}$,

$$\begin{aligned}
&\mathbb{E}\left(\sup_\beta \sum_{i=1}^{n}\left| f_n\left(\beta, \frac{i}{n}\right) - f\left(\beta, \frac{i}{n}\right)\right|\left|U_n\left(\frac{i}{n}\right) - U_n\left(\frac{i-1}{n}\right)\right|\right)^{\alpha-\delta} \\
&\leq \sum_{i=1}^{n}\mathbb{E}\sup_\beta\left| f_n\left(\beta, \frac{i}{n}\right) - f\left(\beta, \frac{i}{n}\right)\right|^{\alpha-\delta}\left|U_n\left(\frac{i}{n}\right) - U_n\left(\frac{i-1}{n}\right)\right|^{\alpha-\delta} \\
&= O\left(\frac{1}{n^{\alpha-\delta}}\right)\sum_{i=1}^{n}\mathbb{E}\left|U_n\left(\frac{i}{n}\right) - U_n\left(\frac{i-1}{n}\right)\right|^{\alpha-\delta} \\
&= O\left(\frac{1}{n^{\alpha-\delta}}\right)\sum_{i=1}^{n}\frac{1}{\left(n^{1/\alpha}L(n)\right)^{\alpha-\delta}} \\
&= O\left(\frac{n}{n^{\alpha-\delta}n^{1-\delta/\alpha}(L(n))^{\alpha-\delta}}\right) = o(1)
\end{aligned}$$

based on the Minkowski adjoint inequality. By Markov's inequality, we can easily obtain

$$\sum_{i=1}^{n} f_n\left(\beta, \frac{i}{n}\right)\frac{X_i}{a_n} - \sum_{i=1}^{n} f\left(\beta, \frac{i}{n}\right)\frac{X_i}{a_n} \xrightarrow{P} 0,$$

on $\mathbb{C}([-M, M])$.

Next, we only consider the weak convergence for $\sum_{i=1}^{n} f\left(\beta, \frac{i}{n}\right) \frac{X_i}{a_n}$. We need to prove the limit relation

$$\frac{1}{a_n}\sum_{i=1}^{n} X_i f\left(\beta, \frac{i}{n}\right) \Rightarrow \sum_{i=1}^{\infty} B_i \Gamma_i^{-1/\alpha} f(\beta, U_i)$$

holds in $\mathbb{C}[0,\infty)$ as $n \to \infty$.

Now pick $\eta > 0$ and define

$$T_\eta : M_p([0,1] \times [-\infty,\infty] \setminus \{0\}) \mapsto \mathbb{C}[0,\infty)$$

in the following way. If

$$m = \sum_j \epsilon_{(t_j,\nu_j)} \in M_p([0,1] \times [-\infty,\infty] \setminus \{0\})$$

and all ν_j's are finite, then

$$(T_\eta m)(\beta) = \sum_k \nu_j 1_{[|\nu_j|>\eta]} f(\beta, t_j).$$

Otherwise, set $(T_\eta m)(\beta) \equiv 0$.

Assume $\beta_n \to \beta$ and $m_n \stackrel{v}{\to} m$ in $M_p([0,1] \times [-\infty,\infty] \setminus \{0\})$, where

$$m\{\partial([0,1] \times \{|\nu| \geq \eta\}) \bigcap [0,1] \times \{-\infty,\infty\}\} = 0$$

implies $(T_\eta m_n)(\beta_n) \to (T_\eta m)(\beta)$. Denote

$$m_n = \sum_j \epsilon_{(t_j^{(n)},\nu_j^{(n)})} \ ; \ m = \sum_j \epsilon_{(t_j,\nu_j)}.$$

The set

$$G_\eta = [0,1] \times \{\nu : |\nu| \geq \eta\}$$

is compact in $[0,1] \times [-\infty,\infty] \setminus \{0\}$ with $m(\partial G_\eta) = 0$. There exist n_0, for $n \geq n_0$,

$$m_n(G_\eta) = m(G_\eta) =: l,$$

say, that there is an enumeration of the points in G_η such that

$$\left((t_k^{(n)}, \nu_k^{(n)}), 1 \leq k \leq l\right) \to ((t_k, \nu_k), 1 \leq k \leq l)$$

and in fact, without loss of generality we may assume for given $\xi > 0$ that

$$\sup_{n \geq n_0} (|\beta_n| \vee \sup_{k=0,\cdots,l} |\nu_k^{(n)}|) \leq \xi.$$

Therefore,

$$\begin{aligned} &|(T_\eta m_n)(\beta_n)| - |(T_\eta m)(\beta)| \\ &= \left| \sum_{k=1}^{l} \nu_k^{(n)} f(\beta_n, t_k^{(n)}) - \sum_{k=1}^{l} \nu_k f(\beta, t_k) \right| \end{aligned}$$

$$\leq \sum_{k=1}^{l} \left| \nu_k^{(n)} f(\beta_n, t_k^{(n)}) - \nu_k f(\beta, t_k) \right|.$$

We have

$$\begin{aligned}
&\left| \nu_k^{(n)} f(\beta_n, t_k^{(n)}) - \nu_k f(\beta, t_k) \right| \\
&= \left| \nu_k^{(n)} f(\beta_n, t_k^{(n)}) - \nu_k f(\beta_n, t_k^{(n)}) + \nu_k f(\beta_n, t_k^{(n)}) - \nu_k f(\beta, t_k) \right| \\
&\leq C \left| \nu_k^{(n)} - \nu_k \right| + |\nu_k| \left| f(\beta_n, t_k^{(n)}) - f(\beta, t_k) \right|.
\end{aligned}$$

Thus

$$\begin{aligned}
&|(T_\eta m_n)(\beta_n)| - |(T_\eta m)(\beta)| \\
&\leq \sum_{k=1}^{l} C \left| \nu_k^{(n)} - \nu_k \right| + \sum_{k=1}^{l} |\nu_k| \left| f(\beta_n, t_k^{(n)}) - f(\beta, t_k) \right| \\
&= \sum_{k=1}^{l} C \left| \nu_k^{(n)} - \nu_k \right| + \sum_{k=1}^{l} |\nu_k| \left| f(\beta_n, t_k^{(n)}) - f(\beta, t_k^{(n)}) + f(\beta, t_k^{(n)}) - f(\beta, t_k) \right| \\
&\leq \sum_{k=1}^{l} C \left| \nu_k^{(n)} - \nu_k \right| + \sum_{k=1}^{l} |\nu_k| \left| f(\beta_n, t_k^{(n)}) - f(\beta, t_k^{(n)}) \right| + \sum_{k=1}^{l} |\nu_k| \left| f(\beta, t_k^{(n)}) - f(\beta, t_k) \right|.
\end{aligned}$$

So

$$\lim_{n\to\infty} |(T_\eta m_n)(\beta_n)| - |(T_\eta m)(\beta)| = 0.$$

This implies that the map $T_\eta : M_p([0,1] \times [-\infty, \infty] \setminus \{0\}) \mapsto \mathbb{C}[0, \infty)$ is continuous a.s. with respect to the distribution of $\mathbb{N}$. Thus we can obtain

$$\begin{aligned}
J_n^\eta(\beta/n) &:= \sum_{j=1}^{n} \frac{X_j}{a_n} f(\beta, \frac{j}{n}) 1_{[|X_j| > \eta_n]} \\
&\Rightarrow \sum_{j=1}^{\infty} B_j \Gamma_j^{-1/\alpha} f(\beta, U_j) 1_{[\Gamma_j^{-1/\alpha} > \eta]} := J_\infty^\eta(\beta)
\end{aligned}$$

in $\mathbb{C}[0, \infty)$.

Also, as $\eta \to 0$ we have

$$J_\infty^\eta(\beta) \Rightarrow J_\infty(\beta) := \sum_{j=1}^{\infty} B_j \Gamma_j^{-1/\alpha} f(\beta, U_j),$$

it remains to prove for any $\theta > 0$,

$$\lim_{\eta\to 0} \limsup_{n\to\infty} \mathbb{P} \left[\left| \left| J_{n,X}^{(\eta)} - J_{n,X} \right| \right| \right] = 0, \tag{2.88}$$

where $||x(\cdot) - y(\cdot)||$ is the $\mathbb{C}[0, \infty)$ metric distance between $x, y \in \mathbb{C}[0, \infty)$. The method of proof of (2.88) will be simply demonstrated if we show for any $\theta > 0$,

$$\lim_{\eta\to 0} \limsup_{n\to\infty} \mathbb{P} \left[\sup_{\beta\in[-M,M]} \left| \sum_{j=1}^{n} \frac{X_j}{a_n} f(\beta, \frac{i}{n}) 1_{[|X_j| < a_n \eta]} \right| > \theta \right] = 0. \tag{2.89}$$

The expression in (2.89) has a bound

$$\begin{aligned}\lim_{\eta\to 0}\limsup_{n\to\infty}\mathbb{P}\left[C\sum_{j=1}^{n}\left|\frac{X_j}{a_n}\right|1_{[|X_j|<a_n\eta]}>\theta\right] &\le \lim_{\eta\to 0}\limsup_{n\to\infty} Cn\mathbb{E}\left(\frac{X_1}{a_n}1_{[|X_1|<a_n\eta]}\right)/\theta,\\ &\le C\lim_{\eta\to 0}\int_{\{x:|x|\le\eta\}}|x|\nu(dx)/\theta = 0\end{aligned}$$

by Karamata's theorem, where

$$\nu(dx) = p\alpha x^{-\alpha-1}dx1_{[x>0]} + q\alpha|x|^{-\alpha-1}dx1_{[x<0]}. \qquad \square$$

Chapter 3

Convergence to Semimartingales

In Chapter 2, the limiting processes for weak convergence are Lévy processes or the product of a Lévy process and an independent random variable (c.f., theorems 2.4 and 2.5). The characteristic functions of these type of processes can be computed. So we can identify limiting processes through the convergence of finite dimensional distributions. When the limiting process is general semimartingale, the finite dimensional distributions are difficult to compute. We need to find new approach. This problem is quite pivotal in some application areas. In some study of physics, chemistry and biology, the diffusion approximation for Markov processes may be very important. In this case, the limiting process of weak convergence may be a diffusion process, which is an important example of semimartingale.

Semimartingale is an adapted stochastic process, which can be used as integrators in the general theory of stochastic integration. Examples of semimartingales include Brownian motion, all local martingales, finite variation processes and Lévy processes. We employ the *predictable characteristics* of semimartingale to discuss the weak convergence of semimartingales. The idea of predictable characteristics of semimartingale comes from the characteristic Lévy triple, which is determined by characteristic functions of finite dimensional distributions of a Lévy process. We list the details of predictable characteristics of a semimartingal in Appendix. The presentation follows Jacod and Shiryaev (2003) closely; any unexplained notation can be found in this monograph.

3.1 The conditions of tightness for semimartingale sequence

In this section, we study a sequence $\{X^n\}$ of $\mathbb{R}$−valued semimartingales, which are defined on the space $(\Omega, \mathcal{F}, (\mathcal{F}_t)_{t\geq 0}, \mathbb{P})$. To simplify, we only discuss the case of $t \in [0,1]$. We wish to derive criteria for tightness of the sequence $\{X^n\}$ in $\mathbb{D}[0,1]$, that can be easily checked.

For càdlàg adapted process sequence $\{X^n\}$, *Aldous (1978)* presented the following criterion for tightness.

Theorem 3.1. *Assume that*

(i) for every $\varepsilon > 0$, there exist $a > 0$ and n_0, such that

$$\mathbb{P}\{\sup_{t\in[0,1]} |X_t^n| \geq a\} \leq \varepsilon$$

as $n \geq n_0$;

(ii) for every $\varepsilon > 0$, $\theta > 0$, and stopping times $S, T \in \mathfrak{S}$,

$$\lim_{\theta\downarrow 0} \limsup_{n\to\infty} \sup_{S\leq T\leq S+\theta} \mathbb{P}(|X_T^n - X_S^n| > \varepsilon) = 0,$$

where $\mathfrak{S}$ is the set of $(\mathcal{F}_t)_{0\leq t\leq 1}-$ stopping times that are bounded by 1.

Then $\{X^n\}$ is tight in $\mathbb{D}[0,1]$.

Proof. It is enough to prove that condition (ii) in Theorem 2.2 holds under condition (ii) in this theorem. Fix $\eta > 0$, condition (ii) implies that for all $\rho > 0$, there are $\delta(\rho) > 0$ and integer $n(\rho) > 0$ such that

$$\mathbb{P}(|X_T^n - X_S^n| \geq \eta) \leq \rho \tag{3.1}$$

for $n \geq n(\rho)$, $\quad S, T \in \mathfrak{S}, \quad S \leq T \leq S + \delta(\rho)$.

Inductively, define stopping times: $S_0^n = 0$,

$$S_{k+1}^n = \inf\{t > S_k^n : |X_t^n - X_{S_k^n}^n| \geq \eta\}.$$

For any given $\varepsilon > 0$, there exist $n_1 = n(\varepsilon)$ and $\delta = \delta(\varepsilon)$, such that

$$\mathbb{P}(S_{k+1}^n \leq (S_k^n + \delta) \wedge 1) \leq \varepsilon,$$

when $n \geq n_1$ by applying (3.1) to $\rho = \varepsilon$ and $S = S_k^n \wedge 1$, $T = S_{k+1}^n \wedge (S_k^n + \delta(\rho)) \wedge 1$, and noticing that

$$|X_{S_k^n}^n - X_{S_{k+1}^n}^n| > \eta$$

if $S_{k+1}^n < \infty$.

Then we choose integer q, such that $q\delta > 2$. The same argument as above shows that there exist $n_2 = n(\varepsilon/q) \vee n_1$ and $\theta = \delta(\varepsilon/q)$, such that

$$\mathbb{P}(S_{k+1}^n \leq (S_k^n + \delta) \wedge 1) \leq \frac{\varepsilon}{q},$$

when $n \geq n_2$.

Since $S_q^n = \sum_{k=1}^q (S_k^n - S_{k-1}^n)$, we have

$$\begin{aligned}
\mathbb{P}(S_q^n < 1) &\geq \mathbb{E}(\sum_{k=1}^q (S_k^n - S_{k-1}^n) 1_{\{S_q^n \leq 1\}}) \\
&\geq \sum_{k=1}^q \mathbb{E}((S_k^n - S_{k-1}^n) 1_{\{S_q^n \leq 1, S_k^n - S_{k-1}^n > \delta\}}) \\
&\geq \sum_{k=1}^q \delta[\mathbb{P}(S_q^n \leq 1) - \mathbb{P}(S_q^n \leq 1, S_k^n - S_{k-1}^n \leq \delta)] \\
&\geq \delta q \mathbb{P}(S_q^n \leq 1) - \delta q \varepsilon.
\end{aligned}$$

We deduce that

$$\mathbb{P}(S_q^n < 1) \leq 2\varepsilon$$

when $n \geq n_2$. Next set

$$A^n = \{S_q^n \geq 1\} \bigcap [\bigcap_{k=1}^{q} \{S_{k+1}^n > (S_k^n + \theta) \wedge 1\}].$$

We can obtain

$$\mathbb{P}(A^n) \geq 1 - 3\varepsilon$$

when $n \geq n_2$.

For $\omega \in A^n$, we consider the subdivision $0 = t_0 < \cdots < t_r = 1$ with $t_i = S_i^n(\omega)$ if $i \leq r-1$ and $r = \inf(i : S_i^n(\omega) \geq 1)$. By the construction of S_j^n's, we have $\omega'(X^n(\omega), \theta) \leq 2\eta$. Thus

$$\mathbb{P}(\omega'(X^n(\omega), \theta) \geq 2\eta) \leq 3\varepsilon$$

when $n \geq n_2$. The proof is completed. □

Next, we consider the case where each X^n is a semimartingale. We present a explicit criteria for tightness of a sequence of semimartingales. We pick a truncation function h, which is bounded and satisfies $h(x) = x$ in a neighborhood of 0. We shall heavily use the predictable characteristics of X^n, $(B^n = B^n(h), C^n, \nu^n)$, as defined in Appendix A. We also introduce the modified second characteristic of X^n as follow:

$$\widetilde{C}^n = \widetilde{C}^n(h) = C^n + h^2 * \nu^n - \sum_{s \leq \cdot} (\Delta B_s^n)^2.$$

Definition 3.1. A sequence $\{X^n\}$ of processes is called $C-$*tight* if it is tight, and if all limiting points of weak convergence of the sequence are laws of continuous processes.

Definition 3.2. For increasing processes X and Y, we say Y *strongly majorizes* X, if $Y - X$ is increasing, denoted by $X \prec Y$.

Lemma 3.1. *The $C-$tightness of B^n and $\widetilde{C}^n$ do not depend on the choice of h.*

Proof. Let h' be another truncation function. There exist two constants $a > 0$, $b > 0$, such that $|h|, |h'| \leq a$, $h(x) = h'(x) = x$ for $|x| \leq b$. Choose positive integer p such that $2/p \leq b$, and denote $g_p(x) = (p|x| - 1)^+ \wedge 1$. Thus

$$|h(x) - h'(x)| \leq 2ag_p(x), \quad (h^2(x) - h'^2(x)) \leq 2a^2 g_p(x),$$

and

$$(h' - h) * \nu^n \prec 2ag_p, \quad (h'^2 - h^2) * \nu^n \prec 2a^2 g_p.$$

Hence $(h'-h)*\nu^n$ and $(h'^2-h^2)*\nu^n$ are $C-$tight. Note that

$$\begin{aligned}&\mathcal{TV}(\sum_{s\le\cdot}[\Delta B^n_s(h)^2-\Delta B^n_s(h')^2])\\&\prec\sum_{s\le\cdot}|\Delta B^n_s(h)^2-\Delta B^n_s(h')^2|[\Delta B^n_s(h)^2+\Delta B^n_s(h')^2]\\&\prec 2a|h-h'|*\nu^n\prec 4a^2(g_p*\nu^n),\end{aligned}$$

where $\mathcal{TV}(A)$ is the totally variation process of A, thus $\sum_{s\le\cdot}[\Delta B^n_s(h)^2-\Delta B^n_s(h')^2]$ is $C-$tight. Since

$$B^n(h')=B^n(h)+(h'-h)*\nu^n,$$

$$\widetilde{C}^n(h')=\widetilde{C}^n(h)+(h^2-h'^2)*\nu^n+\sum_{s\le\cdot}[\Delta B^n_s(h)^2-\Delta B^n_s(h')^2],$$

we can obtain that both B^n and $\widetilde{C}^n$ are $C-$tight by Theorem 1.5. □

Remark 3.1. If Y strongly majorizes X, it is easy to get that the tightness ($C-$tightness) of Y implies the tightness ($C-$tightness) of X.

Theorem 3.2. *Assume that*

(i) the sequence $\{X^n_0\}$ is tight in $\mathbb{R}$;

(ii) for all $\varepsilon>0$,

$$\lim_{a\uparrow\infty}\limsup_{n\to\infty}\mathbb{P}(\nu^n([0,1]\times\{x:|x|>a\})>\varepsilon)=0;\tag{3.2}$$

(iii) $\{B^n\}$, $\{\widetilde{C}^n\}$, $\{g_p*\nu^n\}$ are $C-$tight, where $g_p(x)=(p|x|-1)^+\wedge 1$ for a positive integer p.

Then $\{X^n\}$ is tight in $\mathbb{D}[0,1]$.

Proof. Let h be fixed, and set $h_q=qh(x/q)$. Consider the following decomposition:

$$X^n=U^{nq}+V^{nq}+W^{nq},$$

where $U^{nq}=X^n_0+M^n(h_q)$, $V^{nq}=B^n(h_q)$, $W^{nq}=\check{X}^n(h_q)$. (The definition of $M^n(h_q)$, $B^n(h_q)$, $\check{X}^n(h_q)$ can be found in Appendix A, (3.99).)

From Lemma 3.1, $\{V^{nq}\}$ and $\{\widetilde{C}^n(h_q)\}$ are $C-$tight due to condition (iii). By Proposition 3.2, for all $a>0$ $b>0$,

$$\mathbb{P}(\sup_{t\in[0,1]}|M^n_t(h_q)|\ge a)\le\frac{b}{a^2}+\mathbb{P}(\widetilde{C}^n_1(h_q)\ge b).$$

Thus

$$\begin{aligned}\mathbb{P}(\sup_{t\in[0,1]}|U^{nq}_t|\ge 2a)&\le\mathbb{P}(|X^n_0|>a)+\mathbb{P}(\sup_{t\in[0,1]}|M^n_t(h_q)|\ge a)\\&\le\mathbb{P}(|X^n_0|>a)+\frac{b}{a^2}+\mathbb{P}(\widetilde{C}^n_1(h_q)\ge b).\end{aligned}$$

Furthermore, for stopping times $S, T \in \mathfrak{S}$ with $S \le T$, and for all $\varepsilon > 0$, $\eta > 0$,

$$\begin{aligned}\mathbb{P}(|U_T^{nq} - U_S^{nq}| \ge \varepsilon) &\le \mathbb{P}(\sup_{t\in[0,T]} |U_t^{nq} - U_{t\wedge S}^{nq}| \ge \varepsilon) \\ &\le \frac{\eta}{\varepsilon^2} + \mathbb{P}(\widetilde{C}_T^n(h_q) - \widetilde{C}_S^n(h_q) \ge \eta),\end{aligned}$$

for large enough n. Since $\{X_0^n\}$ is tight and $\{\widetilde{C}^n(h_q)\}$ is C−tight, we can obtain $\{U^{nq}\}$ is tight by Theorem 2.2.

By the definition of h, there is a constant $d > 0$, such that $h_q(x) = x$ when $|x| \le dq$, thus by the definition of $\check{X}^n(h_q)$

$$\begin{aligned}\mathbb{P}(\sup_{t\in[0,1]} |W_t^{nq}| > 0) &\le \mathbb{P}(\sup_{t\in[0,1]} |\Delta X_t| > dq) \\ &\le \varepsilon + \mathbb{P}(\nu^n([0,1] \times \{x : |x| > dq\}) > \varepsilon)\end{aligned}$$

by Proposition 3.2. Then

$$\limsup_{n\to\infty} \mathbb{P}(\sup_{t\in[0,1]} |W_t^{nq}| > 0) = 0$$

by the condition (ii). We obtain this theorem from Theorem 1.5. □

3.2 Weak convergence to semimartingale

This section is the heart of this chapter. We consider the following problem: to find conditions such that a given semimartingale sequence $\{X^n\}$ with predictable characteristics (B^n, C^n, ν^n) and $\widetilde{C}^n$ weakly converges. To simplify the problem, we assume that the candidate limiting process is a canonical process, which is defined on the canonical space $\mathbb{D}[0,1]$. Recall that the canonical space $(\mathbb{D}[0,1], \mathcal{D}[0,1], \mathbf{D}[0,1])$ is the space of all càdlàg functions on $[0,1]$, $\mathcal{D}_t^0[0,1]$ denotes the σ−field generated by all maps: $\beta \to \beta(s)$, $\mathcal{D}_t[0,1] = \bigcap_{t\le u\le 1} \mathcal{D}_u^0[0,1]$, $\mathbf{D}[0,1] = (\mathcal{D}_t[0,1])_{0\le t\le 1}$, $\mathcal{D}[0,1] = \bigvee_{0\le t\le 1} \mathcal{D}_t^0[0,1]$, for $0 \le s \le t \le 1$, and for $\alpha \in \mathbb{D}[0,1]$, canonical process $X_t(\alpha) = \alpha(t)$.

On $(\mathbb{D}[0,1], \mathcal{D}[0,1], \mathbf{D}[0,1])$, the following processes are given:

(i) B is a predictable process with finite variation over finite intervals and $B_0 = 0$;

(ii) C is a continuous adapted process with $C_0 = 0$ and C_t is nondecreasing process in t;

(iii) ν is a predictable random measure on $[0,1] \times \mathbb{R}$, which charges neither $[0,1] \times \{0\}$ nor $\{0\} \times \mathbb{R}$, such that $(1 \wedge |x|^2) * \nu < \infty$,

$$\int_{-\infty}^{+\infty} h(x)\nu(\{t \times dx\}) = \Delta B_t$$

and $\nu(\{t \times \mathbb{R}\}) \le 1$ identically.

We first present the main theorem of this section.

Theorem 3.3. *(Liptser and Shiryaev (1983)) For any dense subset of* $[0,1]$, D, *we assume*

(i) There is a continuous and determinisitic increasing function F_t, such that $Var(B) \prec F$, and $C + (|x|^2 \wedge 1) * \nu \prec F$.

(ii) For all $t \in [0,1]$,

$$\lim_{a\uparrow\infty} \sup_{\alpha\in\mathbb{D}[0,1]} \nu(\alpha : [0,t] \times \{x : |x| > a\}) = 0. \tag{3.3}$$

(iii) There is a unique probability measure $\mathbb{P}'$ on $(\mathbb{D}[0,1], \mathcal{D}[0,1], \mathbf{D}[0,1])$ such that the canonical process X is a semimartingale on $(\mathbb{D}[0,1], \mathcal{D}[0,1], \mathbf{D}[0,1], \mathbb{P}')$ with characteristics (B, C, ν) and initial distribution η.

(iv) For all $t \in D$ and bounded continuous function g, the functions B_t, $\widetilde{C}_t$, $g * \nu_t$ are Skorokhod-continuous on $\mathbb{D}[0,1]$.

(v)

$$\eta^n \xrightarrow{d} \eta, \tag{3.4}$$

where η^n is the initial distribution of X^n.

(vi)

$$\sup_{t\in[0,1]} |B_t^n - B_t \circ X^n| \xrightarrow{\mathbb{P}} 0, \tag{3.5}$$

$$|\widetilde{C}_t^n - \widetilde{C}_t \circ X^n| \xrightarrow{\mathbb{P}} 0 \ \ \textit{for all } t \in D, \tag{3.6}$$

and for all bounded continuous function g

$$|g * \nu_t^n - g * \nu_t \circ X^n| \xrightarrow{\mathbb{P}} 0 \ \ \textit{for all } t \in D. \tag{3.7}$$

Then the laws $\mathcal{L}(X^n)$ weakly converge to $\mathbb{P}'$.

$\mathcal{L}(X^n) \Rightarrow \mathbb{P}'$ implies X^n weakly converge to X. In the following two subsections, we will discuss the conditions in Theorem 3.3.

3.2.1 *Tightness*

In this subsection, our goal is to show the tightness of $\{X^n\}$ if $\{X_0^n\}$ is tight and the conditions (i), (ii), and (vi) in Theorem 3.3 hold.

We verify the conditions in Theorem 3.2.

Let $t \in D$, $g_b(x) = (b|x| - 1)^+ \wedge 1$. For any $\varepsilon > 0$ and $\eta > 0$, there exists $a > 0$ such that

$$g_{2/a} * \nu \leq \frac{\varepsilon}{2}$$

by condition (ii), and

$$\mathbb{P}(|g_{2/a} * \nu_t^n - g_{2/a} * \nu_t \circ X^n| \geq \frac{\varepsilon}{2}) \leq \eta$$

by (3.7). We deduce

$$\mathbb{P}(\nu^n([0,t] \times \{x : |x| > a\}) \geq \varepsilon) \leq \eta$$

since

$$\nu^n([0,t]\times\{x:|x|>a\}) \le g_{2/a} * \nu_t^n.$$

Hence (3.2) holds due to arbitrariness of D.

$\{B\circ X^n\}$ is C–tight, since $Var(B) \prec F$ and F is continuous. From (3.5), $\{B^n\}$ is C–tight.

Let $\varepsilon > 0$, there exist $0 = t_0 < t_1 < \cdots < t_q = 1$ with $t_i \in D$ for $i \ge 0$, such that $F_{t_{i+1}} - F_{t_i} \le \varepsilon$. Then if $t_k \le s \le t_{k+1}$,

$$\begin{aligned}|\widetilde{C}_s^n - \widetilde{C}_s \circ X^n| &\le |\widetilde{C}_s^n - \widetilde{C}_{t_k}^n| + |\widetilde{C}_{t_k}^n - \widetilde{C}_{t_k}\circ X^n| + |\widetilde{C}_{t_k}\circ X^n - \widetilde{C}_s\circ X^n| \\ &\le |\widetilde{C}_{t_{k+1}}^n - \widetilde{C}_{t_k}^n| + |\widetilde{C}_{t_k}^n - \widetilde{C}_{t_k}\circ X^n| + \varepsilon \\ &\le |\widetilde{C}_{t_{k+1}}^n - \widetilde{C}_{t_{k+1}}\circ X^n| + 2|\widetilde{C}_{t_k}^n - \widetilde{C}_{t_k}\circ X^n| + 2\varepsilon.\end{aligned}$$

Thus

$$\sup_{t\in[0,1]} |\widetilde{C}_s^n - \widetilde{C}_s \circ X^n| \le 2\varepsilon + \max_k\{3|\widetilde{C}_{t_k}^n - \widetilde{C}_{t_k}\circ X^n|\}.$$

Then

$$\sup_{t\in[0,1]} |\widetilde{C}_t^n - \widetilde{C}_t \circ X^n| \xrightarrow{\mathbb{P}} 0 \tag{3.8}$$

by (3.6). Similarly,

$$\sup_{t\in[0,1]} |g * \nu_t^n - g * \nu_t \circ X^n| \xrightarrow{\mathbb{P}} 0. \tag{3.9}$$

It is easy to deduce that there exists constant c, such that $\widetilde{C} \prec cF$, $g_b * \nu \prec cF$, thus, $\{\widetilde{C}^n\}$ and $\{g_b * \nu^n\}$ are C–tight. We obtain the tightness of $\{X^n\}$ by Theorem 3.2.

3.2.2 *Identifying the limit*

We first consider the following problem: if a semimartingale sequence $\{X^n\}$ weakly converges to a limiting process X, a semimartingale with predictable characteristics (B, C, ν), are B, C, ν the limits of X^n's predictable characteristics respectively?

This is to say that, we need to clarify the relationship between $\{(B^n, C^n, \nu^n)\}$ and (B, C, ν). The following lemmas are necessary.

Lemma 3.2. *Let H^n be a càdlág adapted process, M^n be a martingale, H, M be càdlág adapted processes defined on the canonical space $(\mathbb{D}[0,1], \mathcal{D}[0,1], \mathbf{D}[0,1])$, and D be a dense subset of $[0,1]$. Assume that*

(i) the family $\{M_t^n\}_{n>0, t\in[0,1]}$ of random variables is uniformly integrable.

(ii) $H^n \Rightarrow H$;

(iii) for all $t \in D$, M_t is $\mathbb{Q}$-a.s. continuous on $\mathbb{D}[0,1]$, where $\mathbb{Q}$ is the distribution of H;

(iv) $M_t^n - M_t \circ H^n \xrightarrow{\mathbb{P}} 0$ for all $t \in D$.

Then the process $M \circ H$ is a martingale with respect to the filtration generated by H.

In general, H can be defined on any given space $(\Omega, \mathcal{F}, \mathbb{P})$. Here, we assume H is the canonical process on the space $(\mathbb{D}[0,1], \mathcal{D}[0,1], \mathbf{D}[0,1], \mathbb{Q})$.

Proof. We first assume that $|M^n| \le b$ identically for some constant $b > 0$. Let $t_1 \le t_2$ with $t_1, t_2 \in D$, and f be a continuous bounded $\mathcal{D}_{t_1-}[0,1]-$measurable function. We first prove that

$$\mathbb{E}_{\mathbb{Q}}[f(H)(M_{t_2} \circ H - M_{t_1} \circ H)] = 0. \tag{3.10}$$

We can deduce that the mappings $\alpha \to f(\alpha)[-b \vee M_{t_i}(\alpha) \wedge b]$, $i = 1, 2$, are $\mathbb{Q} - a.s.$ continuous by (iii), and so

$$\lim_{n\to\infty} \mathbb{E}[f(H^n)(-b \vee (M_{t_i} \circ H^n) \wedge b)] = \mathbb{E}_Q[f(H)(-b \vee (M_{t_i} \circ H) \wedge b)].$$

Furthermore,

$$\lim_{n\to\infty} \mathbb{E}[f(H^n)(-b \vee M^n_{t_i} \wedge b)] = \mathbb{E}_Q[f(H)(-b \vee (M_{t_i} \circ H) \wedge b)] \tag{3.11}$$

since

$$\lim_{n\to\infty} \mathbb{E}[f(H^n)((-b \vee (M_{t_i} \circ H^n) \wedge b) - (-b \vee M^n_{t_i} \wedge b))] = 0$$

by (iv).

For $t \in D$,

$$M^n_t \xrightarrow{d} M_t \circ H$$

by (ii), (iii), (iv). We deduce that $|M \circ H| \le b$ identically outside a $\mathbb{Q}$-null set, since D is dense in $[0,1]$ and M is càdlàg, so $-b \vee (M_{t_i} \circ H) \wedge b = M_{t_i} \circ H$ $\mathbb{Q}$–a.s., and thus

$$\lim_{n\to\infty} \mathbb{E}[f(H^n) M^n_{t_i}] = \mathbb{E}_Q[f(H)(M_{t_i} \circ H)]. \tag{3.12}$$

We can obtain

$$\mathbb{E}_{\mathbb{Q}}[f(H)(M^n_{t_2} - M^n_{t_1})] = 0$$

since M^n is a martingale. Thus (3.10) is obtained by (3.12).

Finally, a monotone class argument shows that (3.10) holds for all $\mathcal{D}_{t_1-}[0,1]-$measurable bounded f. Let $s < t$, there are two sequences $s_n \downarrow s$, $t_n \downarrow t$, such that $s_n, t_n \in D$. So (3.10) holds for every pair (s_n, t_n). By martingale convergence theorem, we can obtain

$$\mathbb{E}_{\mathbb{Q}}[f(H)(M_t \circ H - M_s \circ H)] = 0 \tag{3.13}$$

for all $s \le t$ and $\mathcal{D}_s[0,1]-$measurable bounded f ($\mathcal{D}_s[0,1] \subset \mathcal{D}_{s_n-}[0,1]$). Hence the claim holds when we assume that $|M^n| \le b$ identically for some constant $b > 0$. Set

$$g(b) = \sup_{t\in[0,1], n\ge 1} \mathbb{E}[|M^n_t| - |M^n_t| \wedge b].$$

We have that $g(b) \to 0$ as $b \to \infty$ by the uniformly integrability of the family $\{M^n_t\}_{n>0, t\in[0,1]}$. Moreover,

$$\lim_{n\to\infty} \mathbb{E}[|M^n_t| \wedge b] = \mathbb{E}_{\mathbb{Q}}[|M_t \circ H| \wedge b] \tag{3.14}$$

for $t \in D$ by (3.12). Applying (3.14) for $b' > b$, we deduce

$$\mathbb{E}_{\mathbb{Q}}[|M_t \circ H| \wedge b' - |M_t \circ H| \wedge b] \leq g(b).$$

Let $b' \to \infty$,

$$\mathbb{E}_{\mathbb{Q}}[|M_t \circ H| - |M_t \circ H| \wedge b] \leq g(b). \tag{3.15}$$

On the other hand,

$$\begin{aligned}
&E_{\mathbb{Q}}[|M_t \circ H| 1_{\{|M_t \circ H| > b\}}] \\
&\leq E_{\mathbb{Q}}[(|M_t \circ H| \wedge a) 1_{\{|M_t \circ H| > b\}}] + E_{\mathbb{Q}}[|M_t \circ H| - |M_t \circ H| \wedge a] \\
&\leq a\mathbb{Q}[|M_t \circ H| > b] + g(a) \\
&\leq \sup_{t \in D} \mathbb{E}_Q[|M_t \circ H|] \frac{a}{b} + g(a) \to 0
\end{aligned}$$

under $b \to \infty$, and then $a \to \infty$. Thus the family $(M_t \circ H)_{t \in D}$ is uniformly integrable. It is easy to deduce from (3.11) and (3.15) that (3.12) still holds under the uniformly integrable condition. The same argument will obtain (3.12) holds under the general condition, we obtain this lemma. □

Lemma 3.3. *For semimartingale sequence $\{X^n\}$, we assume that the distribution of X^n on $\mathbb{D}[0,1]$ weakly converges to probability measure $\mathbb{Q}$. Let $\widetilde{D}$ be a dense subset of $[0,1]$, which is contained in $[0,1] \setminus \{t \in [0,1], \mathbb{Q}(\Delta X_t \neq 0) > 0\}$. Moreover,*

(i)

$$|B_t^n - B_t \circ X^n| \xrightarrow{\mathbb{P}} 0 \quad \textit{for all } t \in \widetilde{D}, \tag{3.16}$$

$$|\widetilde{C}_t^n - \widetilde{C}_t \circ X^n| \xrightarrow{\mathbb{P}} 0 \quad \textit{for all } t \in \widetilde{D}, \tag{3.17}$$

and for all bounded continuous function g

$$|g * \nu_t^n - g * \nu_t \circ X^n| \xrightarrow{\mathbb{P}} 0 \quad \textit{for all } t \in \widetilde{D}; \tag{3.18}$$

(ii) $\sup_{\alpha \in \mathbb{D}[0,1]} |\widetilde{C}_t(\alpha)| < \infty$, $\sup_{\alpha \in \mathbb{D}[0,1]} |g * \nu_t(\alpha)| < \infty$ for all $t \in [0,1]$ and every continuous bounded function g;

(iii) for all $t \in [0,1]$ and every continuous bounded function g, the mapping $\alpha \to B_t(\alpha), \alpha \to \widetilde{C}_t(\alpha), \alpha \to g * \nu_t(\alpha)$ are $\mathbb{Q}-$almost surely Skorokhod continuous.

Then X is a semimartingale on $(\mathbb{D}[0,1], \mathcal{D}[0,1], \mathbf{D}[0,1], \mathbb{Q})$ with predictable characteristics (B, C, ν).

Proof. Set

$$X_t' = X_t - \sum_{s \leq t} [\Delta X_s - h(\Delta X_s)], \quad V_t = X_t' - B_t - X_0,$$

$$Z_t = V_t^2 - \widetilde{C}_t, \quad N_t^g = \sum_{s \leq t} g(\Delta X_s) - g * \nu_t,$$

and

$$(X_t^n)' = X_t^n - \sum_{s\le t}[\Delta X_s^n - h(\Delta X_s^n)], \quad V_t^n = (X_t^n)' - B_t^n - X_0^n,$$

$$Z_t^n = (V_t^n)^2 - \widetilde{C}_t^n, \quad N_t^{n,g} = \sum_{s\le t} g(\Delta X_s^n) - g * \nu_t^n.$$

We need to prove that V, Z, N^g are local martingales on $(\mathbb{D}[0,1], \mathcal{D}[0,1], \mathbf{D}[0,1], \mathbb{Q})$. By (ii), we can choose constant K such that

$$|\widetilde{C}_t(\alpha)| + |g * \nu_t(\alpha)| \le K$$

for any fixed bounded continuous function g, and define

$$T_n = \inf\{t : \widetilde{C}_t^n \ge K+1 \ \text{ or } \ g * \nu_t^n \ge K+1\}.$$

We will apply Lemma 3.2 to $H^n = X^n$, $H = X$, $M_t^{1n} = V_{t\wedge T\wedge T_n}^n$, $M_t^1 = V_{t\wedge T}$, $M_t^{2n} = Z_{t\wedge T\wedge T_n}^n$, $M_t^2 = Z_{t\wedge T}$, $M_t^{3n} = N_{t\wedge T\wedge T_n}^{n,g}$, $M_t^3 = N_{t\wedge T}^g$. It is enough to prove that M^1, M^2, M^3 are martingales.

Firstly, by Doob's inequality,

$$\mathbb{E}[\sup_t |M_t^{1n}|] \le 4\mathbb{E}[|M_1^{1n}|^2] = 4\mathbb{E}[\widetilde{C}_{T_n}^n],$$

thus $\{M_t^{1n}\}_{t\in[0,1],n\ge1}$ is uniformly integrable by boundness of $\widetilde{C}_{T_n}^n$.

Similarly, we easily deduce that the family $\{M_t^{2n}\}_{t\in[0,1],n\ge1}$ is uniformly integrable as well. Moreover,

$$\mathbb{E}[\sup_t |M_t^{3n}|] \le 4\mathbb{E}[< M^{3n}, M^{3n} >_{t\wedge T_n}] = 4\mathbb{E}[g^2 * \nu_{t\wedge T_n}^n] \le 4(K+1)^2$$

by Doob's inequality. Thus $\{M_t^{3n}\}_{t\in[0,1],n\ge1}$ is uniformly integrable. So the condition (i) in Lemma 3.2 is satisfied for M^{1n}, M^{2n} and M^{3n}.

One can easily obtain from (iii), M_t^1, M_t^2, M_t^3 are $\mathbb{Q}$-a.s. continuous on $\mathbb{D}[0,1]$ in $\widetilde{D}$.

Note that on $\{T_n \ge T\}$

$$M_t^{1n} - M_t^1 \circ X^n = B_{t\wedge T} \circ X^n - B_{t\wedge T}^n,$$

$$M_t^{3n} - M_t^3 \circ X^n = (g * \nu_{t\wedge T}) \circ X^n - g * \nu_{t\wedge T}^n.$$

(3.17) and (3.18) imply

$$|\widetilde{C}_T^n - \widetilde{C}_T \circ X^n| \xrightarrow{\mathbb{P}} 0,$$

$$|g * \nu_T^n - g * \nu_T \circ X^n| \xrightarrow{\mathbb{P}} 0.$$

Since $\widetilde{C}_T \circ X^n \le K$, $g * \nu_T \circ X^n \le K$, we have

$$\mathbb{P}(\widetilde{C}_T^n \ge K+1) \to 0, \quad \mathbb{P}(g * \nu_T^n \ge K+1) \to 0$$

as $n \uparrow \infty$. Thus

$$\mathbb{P}(T_n < T) \to 0$$

as $n \uparrow \infty$. Then condition (iv) of Lemma 3.2 for M_t^1, M_t^3 is satisfied by (3.16) and (3.18). On the other hand, we have seen above $\{V_{t\wedge T}^n\}$ is uniformly integrable,

$$|V_{t\wedge T\wedge T_n}^n - V_{t\wedge T} \circ X^n| \xrightarrow{\mathbb{P}} 0$$

and

$$\begin{aligned} M_t^{2n} - M_t^2 \circ X^n = & V_{t\wedge T\wedge T_n}^n (V_{t\wedge T\wedge T_n}^n - V_{t\wedge T} \circ X^n) \\ & + V_{t\wedge T} \circ X^n (V_{t\wedge T\wedge T_n}^n - V_{t\wedge T} \circ X^n) + \widetilde{C}_{t\wedge T} \circ X^n - \widetilde{C}_{t\wedge T\wedge T_n}^n. \end{aligned}$$

Then, condition (iv) of Lemma 3.2 for M_t^2 is satisfied. We obtain this lemma. □

Lemma 3.3 means that probability measure $\mathbb{Q}$ is a solution of martingale problem (B, C, ν). When we assume that the limiting process is the canonical process X, the distribution of X in $\mathbb{D}[0,1]$ is a probability measure on $\mathbb{D}[0,1]$. If the martingale problem has unique solution on $\mathbb{D}[0,1]$, it means that the distribution of X in $\mathbb{D}[0,1]$, the measure on $\mathbb{D}[0,1]$, under the tightness assumption is unique, thus we identify the limiting process.

The proof of Theorem 3.3. According to previous discussion and the condition (iii), it remains to prove that $D' = D \bigcap \{[0,1] \setminus \{t > 0, \mathbb{P}'(\Delta X_t \neq 0) > 0\}\}$ is dense in $[0,1]$.

For any $t \in [0,1]$ and the positive rational number ε, there exist $s, s' \in D$ with $s < t < s'$, and

$$g_{2/\varepsilon} * \nu_{s'} - g_{2/\varepsilon} * \nu_s \leq \varepsilon$$

since condition (i). By Proposition 1.10, there also exist $r, r' \in [0,1] \setminus \{t > 0, \mathbb{P}'(\Delta X_t \neq 0) > 0\}$ with $s \leq r < t < r' \leq s'$,

$$\sup_{r<u<r'} |\Delta X_u^n| \xrightarrow{d} \sup_{r<u<r'} |\Delta X_u|.$$

Note that

$$\begin{aligned} \mathbb{P}'(|\Delta X_t| \geq \varepsilon) &\leq \mathbb{P}'(\sup_{r<u<r'} |\Delta X_u| \geq \varepsilon) \\ &\leq \limsup_{n\to\infty} \mathbb{P}(\sup_{r<u<r'} |\Delta X_u^n| \geq \varepsilon) \\ &\leq \limsup_{n\to\infty} \mathbb{P}(\sup_{s<u<s'} |\Delta X_u^n| \geq \varepsilon) \\ &\leq \limsup_{n\to\infty} \mathbb{P}(\sum_{s<u<s'} g_{2/\varepsilon}(\Delta X_u^n) \geq 1). \end{aligned}$$

Then, by Proposition 3.2,

$$\mathbb{P}'(|\Delta X_t| \geq \varepsilon) \leq 2\varepsilon + \limsup_{n\to\infty} \mathbb{P}(g_{2/\varepsilon} * \nu_{s'}^n - g_{2/\varepsilon} * \nu_s^n \geq 2\varepsilon)$$

since $1_{(s,\infty)} g_{2/\varepsilon} * \nu^n$ is the $\mathbb{P}-$compensator of $\sum_{s<u<s'} g_{2/\varepsilon}(\Delta X_u^n)$.

At last,

$$|g_{2/\varepsilon} * \nu_s^n - g_{2/\varepsilon} * \nu_s \circ X^n| \xrightarrow{\mathbb{P}} 0$$

implies

$$\mathbb{P}'(|\Delta X_t| > \varepsilon) \leq 2\varepsilon$$

for any t. Thus, $\mathbb{P}'(|\Delta X_t| > 0) = 0$ since ε is arbitrary. We obtain

$$\{t > 0, \mathbb{P}'(\Delta X_t \neq 0) > 0\} = \emptyset.$$

The proof is complete.

3.3 Weak convergence to stochastic integral I: the martingale convergence approach

Itô type stochastic integral is an important example of semimartingale. Weak convergence to Itô type stochastic integral is an interesting problem in probability and statistics. In the following two sections, we will present some results on convergence to stochastic integral. In this section, we obtain the result by means of Theorem 3.3, this technique is called the martingale convergence approach.

Firstly, we consider the weak convergence of functionals for sums of *causal processes.* We call $\{X_n, n \geq 1\}$ a causal process if X_n has the form

$$X_n = g(\cdots, \varepsilon_{n-1}, \varepsilon_n),$$

where $\{\varepsilon_n; n \in Z\}$ is mean zero, i.i.d. random variables and g is a measurable function. Causal process is a very important example of stationary processes. It has been widely used in practice, and contains many important statistical models, such as linear processes, ARCH models, threshold AR (TAR) models and so on. Asymptotic behavior of the sums of causal processes, $S_n = \sum_{i=1}^n X_i$, is important subjects in both practice and theory.

Recall that $Z \in \mathcal{L}^p$ $(p > 0)$ if $||Z||_p = [\mathbb{E}(|Z|^p)]^{1/p} < \infty$ and write $||Z|| = ||Z||_2$.

To study the asymptotic property of the sums of causal processes, *martingale approximation* is an effective method. Roughly speaking, martingale approximation is to find a martingale M_n, such that the error $\| S_n - M_n \|_p$ is small in some sense.

Define

$$\mathcal{P}_k Z = \mathbb{E}(Z|\mathcal{F}_k) - \mathbb{E}(Z|\mathcal{F}_{k-1}), \qquad Z \in \mathcal{L}^1,$$

$$D_k = \sum_{i=k}^{\infty} \mathcal{P}_k X_i, \qquad M_k = \sum_{i=1}^{k} D_i \qquad R_k = S_k - M_k,$$

$$H_k = \sum_{i=1}^{\infty} \mathbb{E}(X_{k+i}|\mathcal{F}_k),$$

$$\theta_{n,p} = ||\mathcal{P}_0 X_n||_p, \qquad \Lambda_{n,q} = \sum_{i=0}^{n} \theta_{i,q},\ n > 0,$$

$$\Theta_{m,p} = \sum_{i=m}^{\infty} \theta_{i,p}, \qquad \theta_{n,p} = 0 = \Lambda_{n,p} \text{ if } n < 0.$$

It is easy to check that $\{D_k\}$ is a sequence of martingale differences, M_k is a martingale. We will use M_k to approximate S_k, and assume that D_k converges almost surely.

We present the following theorem.

Theorem 3.4. *(Lin and Wang (2010)) Let $f : \mathbb{R} \to \mathbb{R}$ be a twice continuously differentiable function satisfying $|f'(x)| \leq C(1 + |x|^{\alpha})$ for some positive constants C, α and all $x \in \mathbb{R}$. Suppose that X_t is a causal process satisfying*

(i) $X_0 \in \mathcal{L}^q$, $q > 2$, and $\Theta_{n,q^*} = O(n^{1/q^*-1/2}(\log n)^{-1})$, where $q^* = \min(q, 4)$;

(ii)

$$\sum_{k=1}^{\infty} ||\mathbb{E}(D_k^2|\mathcal{F}_0) - \sigma^2||_{q^*/2} < \infty,$$

where $||D_k|| = \sigma$;

(iii)

$$\sum_{k=0}^{\infty}\sum_{i=1}^{\infty} ||\mathbb{E}(X_k X_{k+i}|\mathcal{F}_0) - \mathbb{E}(X_k X_{k+i}|\mathcal{F}_{-1})||_4 < \infty$$

and

$$||\sum_{k=0}^{\infty}\sum_{i=1}^{\infty} \mathbb{E}(X_k X_{k+i}|\mathcal{F}_0)||_3 < \infty,$$

where $\mathcal{F}_k = (\cdots, \varepsilon_{k-1}, \varepsilon_k)$. Then

$$\frac{1}{\sqrt{n}} \sum_{t=2}^{[n\cdot]} f\left(\frac{1}{\sqrt{n}} \sum_{i=1}^{t-1} X_i\right) X_t \Rightarrow \lambda \int_0^{\cdot} f'(W(v))dv + \sigma \int_0^{\cdot} f(W(v))dW(v) \tag{3.19}$$

where $\lambda = \sum_{j=1}^{\infty} \mathbb{E}X_0 X_j$, W is the standard Brownian motion.

To prove Theorem 3.4, we verify the conditions in Theorem 3.3. Firstly, we introduce two lemmas.

Lemma 3.4. *(Wu (2007)) Assume that $\mathbb{E}[X_0] = 0$, $X_0 \in \mathcal{L}^q$, $q > 1$, and $\Theta_{0,q} = \sum_{i=0}^{\infty} \theta_{i,q} < \infty$, then*

$$\| \max_{k \leq n} |S_k| \|_q \leq \frac{qB_q}{q-1} n^{1/q'} \Theta_{0,q},$$

where $q' = \min(2, q)$, $B_q = 18q^{3/2}(q-1)^{-1/2}$ if $q \in (1, 2) \cup (2, \infty)$ and $B_q = 1$ if $q = 2$.

Lemma 3.5. *(Wu (2007)) Under conditions (i) and (ii), there exists a standard Brownian motion W on a richer probability space such that*

$$|S_n - \sigma W(n)| = O_{a.s.}(n^{1/4}(\log n)^{1/2}(\log\log n)^{1/4}).$$

Set

$$X_n(s) = (\frac{1}{\sqrt{n}}\sum_{t=2}^{[ns]} f(\frac{1}{\sqrt{n}}\sum_{i=1}^{t-1} X_i)X_t, \frac{1}{\sqrt{n}}\sum_{t=1}^{[ns]} X_t) =: (X_n^1(s), X_n^2(s))$$

and

$$X(s) = (\lambda \int_0^s f'(B(v))dv + \sigma \int_0^s f(B(v))dB(v), B(s)) =: (X^1(s), X^2(s)).$$

We can get the first two predictable characteristics of X_n as follows:

$$B_n(s) = (\frac{1}{\sqrt{n}}\sum_{t=2}^{[ns]} f(\frac{1}{\sqrt{n}}\sum_{i=1}^{t-1} X_i)(X_t - D_t), \frac{1}{\sqrt{n}}\sum_{t=1}^{[ns]}(X_t - D_t)),$$

$$C_n(s) = \begin{bmatrix} C_n^{11}(s) & C_n^{12}(s) \\ C_n^{21}(s) & C_n^{22}(s) \end{bmatrix},$$

$$C_n^{11}(s) = \frac{1}{n}\sum_{t=2}^{[ns]} f^2(\frac{1}{\sqrt{n}}\sum_{i=1}^{t-1} X_i)E(D_t^2|\mathcal{F}_{t-1}),$$

$$C_n^{22}(s) = \frac{1}{n}\sum_{t=1}^{[ns]} E(D_t^2|\mathcal{F}_{t-1}),$$

$$C_n^{12}(s) = C_n^{21}(s) = \frac{1}{n}\sum_{t=2}^{[ns]} f(\frac{1}{\sqrt{n}}\sum_{i=1}^{t-1} X_i)E(D_t^2|\mathcal{F}_{t-1}).$$

The process $X(s) = (X^1(s), X^2(s))$, a random element in the Skorokhod space $\mathbb{D}([0,1]\times[0,1])$, is a solution to the stochastic differential equation

$$\begin{cases} dX^1(t) = \lambda f'(X^2(t))dt + \sigma f(X^2(t))dW(t), \\ dX^2(t) = dW(t). \end{cases} \tag{3.20}$$

The predictable characteristics of X are $(B(X), C(X), 0)$:

$$B(s,\alpha) = (\lambda \int_0^s f'(\sigma\alpha_2(v))dv, 0),$$

$$C(s,\alpha) = \begin{bmatrix} \sigma^2 \int_0^s f^2(\alpha_2(v))dv & \sigma \int_0^s f(\alpha_2(v))dv \\ \sigma \int_0^s f(\alpha_2(v))dv & \sigma^2 s \end{bmatrix}.$$

The proof of Theorem 3.4. We can easily obtain

$$B(s)\circ X_n = \frac{\lambda}{n}\sum_{t=2}^{[ns]} f'(\frac{1}{\sqrt{n}}\sum_{i=1}^{t-1} X_i) + \frac{\lambda}{n} f'(\frac{1}{\sqrt{n}}\sum_{i=1}^{[ns]} X_i)(ns - [ns]),$$

$$C^{11}(s)\circ X_n = \frac{\sigma^2}{n}\sum_{t=2}^{[ns]} f^2(\frac{1}{\sqrt{n}}\sum_{i=1}^{t-1} X_i) + \frac{\sigma^2}{n} f^2(\frac{1}{\sqrt{n}}\sum_{i=1}^{[ns]} X_i)(ns-[ns]),$$

$$C^{12}(s)\circ X_n = C^{21}(s)\circ X_n = \frac{\sigma^2}{n}\sum_{t=2}^{[ns]} f(\frac{1}{\sqrt{n}}\sum_{i=1}^{t-1} X_i) + \frac{\sigma^2}{n} f(\frac{1}{\sqrt{n}}\sum_{i=1}^{[ns]} X_i)(ns-[ns]).$$

Firstly, we need to prove

$$\sup_{0<s\le 1} |B_n(s) - B(s)\circ X_n| \xrightarrow{\mathbb{P}} 0, \tag{3.21}$$

which will be proved if we show

$$\sup_{0<s\le 1} |\frac{1}{\sqrt{n}}\sum_{t=2}^{[ns]} f(\frac{1}{\sqrt{n}}\sum_{i=1}^{t-1} X_i)(X_t - D_t) - \frac{\lambda}{n}\sum_{t=2}^{[ns]} f'(\frac{1}{\sqrt{n}}\sum_{i=1}^{t-1} X_i)| \xrightarrow{\mathbb{P}} 0. \tag{3.22}$$

We have

$$\begin{aligned}
\mathcal{J}_k &:= |\frac{1}{\sqrt{n}}\sum_{t=2}^{k} f(\frac{1}{\sqrt{n}}\sum_{i=1}^{t-1} X_i)(X_t - D_t) - \frac{\lambda}{n}\sum_{t=2}^{k} f'(\frac{1}{\sqrt{n}}\sum_{i=1}^{t-1} X_i)| \\
&= |\frac{1}{\sqrt{n}}\sum_{t=2}^{k} f(\frac{1}{\sqrt{n}}\sum_{i=1}^{t-1} X_i)(H_{t-1} - H_t) - \frac{\lambda}{n}\sum_{t=2}^{k} f'(\frac{1}{\sqrt{n}}\sum_{i=1}^{t-1} X_i)| \\
&= |\frac{1}{\sqrt{n}}\sum_{t=2}^{k}(f(\frac{1}{\sqrt{n}}\sum_{i=1}^{t} X_i) - f(\frac{1}{\sqrt{n}}\sum_{i=1}^{t-1} X_i))H_t \\
&\quad -\frac{1}{\sqrt{n}} f(\frac{1}{\sqrt{n}}\sum_{i=1}^{k} X_i)H_k - \frac{\lambda}{n}\sum_{t=2}^{k} f'(\frac{1}{\sqrt{n}}\sum_{i=1}^{t-1} X_i)|
\end{aligned}$$

and

$$\begin{aligned}
\max_{1\le k\le n} \mathcal{J}_k \le{}& \max_{1\le k\le n} |\frac{1}{\sqrt{n}} f(\frac{1}{\sqrt{n}}\sum_{i=1}^{k} X_i)H_k| + \max_{1\le k\le n} |\frac{1}{n}\sum_{t=2}^{k} f'(\frac{1}{\sqrt{n}}\sum_{i=1}^{t-1} X_i)(X_t H_t - \lambda)| \\
&+ \max_{1\le k\le n} |\frac{1}{\sqrt{n}}\sum_{t=2}^{k}(f(\frac{1}{\sqrt{n}}\sum_{i=1}^{t} X_i) - f(\frac{1}{\sqrt{n}}\sum_{i=1}^{t-1} X_i) - f'(\frac{1}{\sqrt{n}}\sum_{i=1}^{t-1} X_i)\frac{X_t}{\sqrt{n}})H_t|.
\end{aligned}$$

Firstly, by the definition of H_k,

$$\max_{1\le k\le n} |\frac{1}{\sqrt{n}} f(\frac{1}{\sqrt{n}}\sum_{i=1}^{k} X_i)H_k| \xrightarrow{\mathbb{P}} 0. \tag{3.23}$$

Next, we prove

$$\max_{1\le k\le n} |\frac{1}{n}\sum_{t=2}^{k} f'(\frac{1}{\sqrt{n}}\sum_{i=1}^{t-1} X_i)(X_t H_t - \lambda)| \xrightarrow{\mathbb{P}} 0. \tag{3.24}$$

Set $Y_{t,j} = \mathbb{E}(X_t X_{t+j}|\mathcal{F}_t) - \mathbb{E}(X_t X_{t+j})$, (3.24) can be implied by

$$\max_{1\le k\le n} |\frac{1}{n}\sum_{t=2}^{k} f'(\frac{1}{\sqrt{n}}\sum_{i=1}^{t-1} X_i)(\sum_{j=1}^{\infty} Y_{t,j})| \xrightarrow{\mathbb{P}} 0. \tag{3.25}$$

We approximate $S_t := \sum_{j=1}^{\infty} Y_{t,j}$ by $\widetilde{D}_t := \sum_{k=t}^{\infty} \mathcal{P}_t(S_k)$, then we need to prove

$$\max_{1\le k\le n} |\frac{1}{n}\sum_{t=2}^{k} f'(\frac{1}{\sqrt{n}}\sum_{i=1}^{t-1} X_i)\widetilde{D}_t| \xrightarrow{\mathbb{P}} 0 \tag{3.26}$$

and

$$\max_{1\le k\le n} |\frac{1}{n}\sum_{t=2}^{k} f'(\frac{1}{\sqrt{n}}\sum_{i=1}^{t-1} X_i)(S_t - \widetilde{D}_t)| \xrightarrow{\mathbb{P}} 0. \tag{3.27}$$

For (3.26), we have

$$\mathbb{E}(f'(\frac{1}{\sqrt{n}}\sum_{i=1}^{t-1} X_i)\widetilde{D}_t)^2 \le \sqrt{E(f'(\frac{1}{\sqrt{n}}\sum_{i=1}^{t-1} X_i)^4)E(\widetilde{D}_t)^4},$$

$$\begin{aligned}
[\mathbb{E}(\widetilde{D}_t)^4]^{1/4} &= [\mathbb{E}(\sum_{k=t}^{\infty} \mathcal{P}_t(S_k))^4]^{1/4} \\
&\le \sum_{k=t}^{\infty} ||P_t(S_k)||_4 = \sum_{k=t}^{\infty}\sum_{i=1}^{\infty} ||\mathbb{E}(X_k X_{k+i}|\mathcal{F}_t) - \mathbb{E}(X_k X_{k+i}|\mathcal{F}_{t-1})||_4 < \infty
\end{aligned}$$

by condition (iii). Since $f'(x) \le C(1+|x|^\alpha)$, we have

$$\mathbb{E}(f'(\frac{1}{\sqrt{n}}\sum_{i=1}^{t-1} X_i)\widetilde{D}_t)^2 \le L \tag{3.28}$$

by Lemma 3.4. Then, by the Kolmogorov inequality for martingale, for any $\varepsilon > 0$ we have

$$\begin{aligned}
&\mathbb{P}(\max_{1\le k\le n} |\frac{1}{n}\sum_{t=2}^{k} f'(\frac{1}{\sqrt{n}}\sum_{i=1}^{t-1} X_i)\widetilde{D}_t| > \varepsilon) \\
&\le \frac{\mathbb{E}[\sum_{t=2}^{n} f'(\frac{1}{\sqrt{n}}\sum_{i=1}^{t-1} X_i)\widetilde{D}_t]^2}{n^2\varepsilon^2} \\
&\le \frac{\max_{2\le t\le n} \mathbb{E}[f'(\frac{1}{\sqrt{n}}\sum_{i=1}^{t-1} X_i)\widetilde{D}_t]^2}{n\varepsilon^2} = O(\frac{1}{n}),
\end{aligned}$$

(3.26) is proved.

For (3.27), we have

$$S_t - \widetilde{D}_t = Z_{t-1} - Z_t, \quad Z_t = \sum_{i=1}^{\infty}\sum_{k=1}^{\infty} \mathbb{E}(X_{t+i+k}X_{t+i}|\mathcal{F}_t)$$

and

$$\begin{aligned}
&\max_{1\le k\le n} |\frac{1}{n}\sum_{t=2}^{k} f'(\frac{1}{\sqrt{n}}\sum_{i=1}^{t-1} X_i)(S_t - \widetilde{D}_t)| = \max_{1\le k\le n} |\frac{1}{n}\sum_{t=2}^{k} f'(\frac{1}{\sqrt{n}}\sum_{i=1}^{t-1} X_i)(Z_t - Z_{t-1})| \\
&\le \max_{1\le k\le n} |\frac{1}{n} f'(\frac{1}{\sqrt{n}}\sum_{i=1}^{k} X_i)Z_k| + \max_{1\le k\le n} |\frac{1}{n}\sum_{t=2}^{k}(f'(\frac{1}{\sqrt{n}}\sum_{i=1}^{t} X_i) - f'(\frac{1}{\sqrt{n}}\sum_{i=1}^{t-1} X_i))Z_t|
\end{aligned}$$

$$\leq \max_{1\leq k\leq n}|f'(\frac{1}{\sqrt{n}}\sum_{i=1}^{k}X_i)|\max_{1\leq k\leq n}\frac{|Z_k|}{n}+\max_{1\leq k\leq n}\frac{|X_kZ_k|}{\sqrt{n}}\sup_{|t|\leq\max_{1\leq k\leq n}|\frac{\sum_{i=1}^{k}X_i}{\sqrt{n}}|}f''(t).$$

Under condition (iii), by the law of large numbers, we have

$$\max_{1\leq k\leq n}n^{-\frac{1}{6}}|X_k|\xrightarrow{\mathbb{P}}0,\quad \max_{1\leq k\leq n}n^{-\frac{1}{3}}|Z_k|\xrightarrow{\mathbb{P}}0,$$

so

$$\max_{1\leq k\leq n}\frac{1}{\sqrt{n}}|X_kZ_k|\xrightarrow{\mathbb{P}}0.$$

From Lemma 3.5, we have $\frac{1}{\sqrt{n}}\sum_{i=1}^{k}X_i=O_{\mathbb{P}}(1)$. By the continuity of $f''(x)$, we obtain (3.27), thus (3.25) and further (3.24) hold.

We have, by the Taylor expansion, that

$$\max_{1\leq k\leq n}|\frac{1}{\sqrt{n}}\sum_{t=2}^{k}(f(\frac{1}{\sqrt{n}}\sum_{i=1}^{t}X_i)-f(\frac{1}{\sqrt{n}}\sum_{i=1}^{t-1}X_i)-f'(\frac{1}{\sqrt{n}}\sum_{i=1}^{t-1}X_i)\frac{X_t}{\sqrt{n}})H_t|$$
$$\leq\frac{1}{2}\max_{1\leq k\leq n}\frac{1}{\sqrt{n}}X_k^2|H_k|\sup_{|t|\leq\max_{1\leq k\leq n}|\frac{\sum_{i=1}^{k}X_i}{\sqrt{n}}|}f''(t).$$

Under condition (iii), by the law of large numbers and similar argument in the above, we get that

$$\max_{1\leq k\leq n}|\frac{1}{\sqrt{n}}\sum_{t=2}^{k}(f(\frac{1}{\sqrt{n}}\sum_{i=1}^{t}X_i)-f(\frac{1}{\sqrt{n}}\sum_{i=1}^{t-1}X_i)-f'(\frac{1}{\sqrt{n}}\sum_{i=1}^{t-1}X_i)\frac{X_t}{\sqrt{n}})H_t|\xrightarrow{\mathbb{P}}0.$$

Then combining it with (3.23) and (3.24), we obtain

$$\max_{1\leq k\leq n}\mathcal{J}_k\xrightarrow{\mathbb{P}}0,$$

which implies (3.22).

Next, we prove

$$\sup_{0<s\leq 1}|C_n^{ij}(s)-C^{ij}(s)\circ X_n|\xrightarrow{\mathbb{P}}0. \tag{3.29}$$

We only consider the case of $i=j=1$, since the proofs for other cases are similar. Clearly, (3.29) is equivalent to

$$\sup_{0<s\leq 1}|\frac{1}{n}\sum_{t=2}^{[ns]}f^2(\frac{1}{\sqrt{n}}\sum_{i=1}^{t-1}X_i)(E(D_t^2|\mathcal{F}_{t-1})-\sigma^2)-\frac{\sigma^2}{n}f^2(\frac{1}{\sqrt{n}}\sum_{i=1}^{[ns]}X_i)(ns-[ns])|\xrightarrow{\mathbb{P}}0,$$

Firstly, we prove

$$\sup_{0<s\leq 1}|\frac{1}{n}\sum_{t=2}^{[ns]}f^2(\frac{1}{\sqrt{n}}\sum_{i=1}^{t-1}X_i)\sigma^2-\frac{1}{n}\sum_{t=2}^{[ns]}f^2(\frac{1}{\sqrt{n}}\sum_{i=1}^{t-1}D_i)\sigma^2|\xrightarrow{\mathbb{P}}0. \tag{3.30}$$

Since

$$\sup_{0<s\le 1}|\frac{1}{n}\sum_{t=2}^{[ns]} f^2(\frac{1}{\sqrt{n}}\sum_{i=1}^{t-1} X_i) - \frac{1}{n}\sum_{t=2}^{[ns]} f^2(\frac{1}{\sqrt{n}}\sum_{i=1}^{t-1} D_i)|$$
$$\le \max_{1\le t\le n} |f^2(\frac{1}{\sqrt{n}}\sum_{i=1}^{t-1} X_i) - f^2(\frac{1}{\sqrt{n}}\sum_{i=1}^{t-1} D_i)|,$$

and f is a uniform continuous function, we can get (3.30) by

$$\max_{1\le t\le n}|\frac{1}{\sqrt{n}}\sum_{i=1}^{t-1} X_i - \frac{1}{\sqrt{n}}\sum_{i=1}^{t-1} D_i| \xrightarrow{\mathbb{P}} 0. \tag{3.31}$$

For any $\varepsilon > 0$, by Lemma 3.4, for $2 < q < 4$, we have

$$\mathbb{P}(\frac{1}{\sqrt{n}}\max_{1\le t\le n}|\sum_{i=1}^{t-1} X_i - \sum_{i=1}^{t-1} D_i| > \varepsilon) \le \frac{\mathbb{E}[\sum_{t=1}^{n}(X_t - D_t)]^2}{n\varepsilon^2} \le C\frac{n^{1/q'}(\log n)^{-1}}{n\varepsilon^2},$$

which implies (3.31), then we obtain (3.30).

By the martingale version of the *Skorokhod representation theorem*, on a richer probability space, there exist a standard Brownian motion W and nonnegative random variables $\tau_1, \tau_2, \cdots$ with partial sums $T_k = \sum_{i=1}^{k}\tau_i$ such that $T_k - k\sigma^2 = o_{\text{a.s.}}(k^{2/q})$ and $M_k = W(T_k)$, $\mathbb{E}(\tau_k|\mathcal{F}_{k-1}) = \mathbb{E}(D_k^2|\mathcal{F}_{k-1})$ for $k \ge 1$. (cf. *Strassen (1964)*).

For $\frac{T_{k-1}}{n} < s \le \frac{T_k}{n}$, we consider

$$\mathcal{I}_n(s) = \frac{\sigma^2}{n}\sum_{t=2}^{[ns]} f^2(\frac{1}{\sqrt{n}}\sum_{i=1}^{t-1} D_i) + \frac{\sigma^2}{n} f^2(\frac{1}{\sqrt{n}}\sum_{i=1}^{[ns]} D_i)(ns - [ns]).$$

By the martingale version of the Skorokhod representation theorem, we have

$$\mathcal{I}_n(s) = \frac{\sigma^2}{n}\sum_{t=2}^{k-1} f^2(W(\frac{T_{t-1}}{n})) + \frac{\sigma^2}{n} f^2(W(\frac{T_{k-1}}{n}))(ns - [ns]).$$

Since $T_k - k\sigma^2 = o_{\text{a.s.}}(k^{2/q})$, by using the continuity modulas theorem for the Wiener process

$$\max_{t\le k}|W(\frac{T_t}{n}) - W(\frac{\sigma^2 t}{n})| \le \max_{t\le k}\sup_{|x-\sigma^2 t|\le k^{2/q}}|W(\frac{x}{n}) - W(\frac{\sigma^2 t}{n})|$$
$$\le o_{\text{a.s.}}(k^{1/q}\sqrt{\log k}).$$

By the similar argument in (3.30), we have

$$\sup_{0<s\le 1}|\mathcal{I}_n(s) - \sum_{t=2}^{k-1} f^2(W(\frac{\sigma^2(t-1)}{n}))\frac{\sigma^2}{n} + \frac{\sigma^2}{n} f^2(W(\frac{\sigma^2(k-1)}{n}))(ns - [ns])| \xrightarrow{\mathbb{P}} 0.$$

By the Riemann approximation of stochastic integral, and the continuity of Brownian motion's paths, we have

$$\sup_{0<s\le 1}|\mathcal{I}_n(s) - \int_0^s f^2(W(v))dv| \xrightarrow{\mathbb{P}} 0. \tag{3.32}$$

By noting $M_k = W(T_k)$, we have

$$\begin{aligned}
&\sup_{0<s\leq 1} |\frac{1}{n}\sum_{t=2}^{[ns]} f^2(\frac{1}{\sqrt{n}}\sum_{i=1}^{t-1} X_i)\mathbb{E}(D_t^2|\mathcal{F}_{t-1}) - \frac{1}{n}\sum_{t=2}^{[ns]} f^2(\frac{1}{\sqrt{n}}W(T_{t-1}))\mathbb{E}(D_t^2|\mathcal{F}_{t-1})| \\
&= \sup_{0<s\leq 1} |\frac{1}{n}\sum_{t=2}^{[ns]} f^2(\frac{1}{\sqrt{n}}\sum_{i=1}^{t-1} X_i)\mathbb{E}(T_t - T_{t-1}|\mathcal{F}_{t-1}) \\
&\quad - \frac{1}{n}\sum_{t=2}^{[ns]} f^2(\frac{1}{\sqrt{n}}W(T_{t-1}))\mathbb{E}(T_t - T_{t-1}|\mathcal{F}_{t-1})| \\
&\xrightarrow{\mathbb{P}} 0
\end{aligned}$$

by Lemma 3.5.

By the Riemann approximation of stochastic integral again and the Approximated Laplacians property (cf. *Dellacherie and Meyer (1982)*), we obtain

$$\sup_{0<s\leq 1} |\frac{1}{n}\sum_{t=2}^{[ns]} f^2(\frac{1}{\sqrt{n}}W(T_{t-1}))\mathbb{E}(T_t - T_{t-1}|\mathcal{F}_{t-1}) - \int_0^s f^2(W(v))dv| \xrightarrow{\mathbb{P}} 0,$$

thus, we have

$$\sup_{0<s\leq 1} |\frac{1}{n}\sum_{t=2}^{[ns]} f^2(\frac{1}{\sqrt{n}}\sum_{i=1}^{t-1} X_i)\mathbb{E}(T_t - T_{t-1}|\mathcal{F}_{t-1}) - \int_0^s f^2(W(v))dv| \xrightarrow{\mathbb{P}} 0. \tag{3.33}$$

By (3.30), (3.32) and (3.33), we obtain (3.29).

Furthermore, we prove

$$\sup_{0<s\leq 1} |\Delta X_n(s)| \xrightarrow{\mathbb{P}} 0. \tag{3.34}$$

In fact,

$$\sup_{0<s\leq 1} |\Delta X_n(s)| \leq \max_{0\leq k\leq n} |f(\frac{1}{\sqrt{n}}\sum_{i=1}^{k} X_i)| \cdot \max_{0\leq k\leq n} \frac{1}{\sqrt{n}}|X_k| + \max_{0\leq k\leq n} \frac{1}{\sqrt{n}}|X_k|. \tag{3.35}$$

From Lemma 3.5, we have $\frac{1}{\sqrt{n}}\sum_{i=1}^{k} X_i = O_{\mathbb{P}}(1)$. Combining it with the assumptions for $f(x)$, we obtain (3.34) by (3.35).

(3.34) implies (3.7), since the limiting process is continuous.

Under the assumptions of the theorem, functions $f(x)$ and $f'(x)$ are locally Lipschitz continuous and satisfy the growth condition (i.e. $|f(x)| \leq C(1+|x|^\alpha)$ for $\alpha > 0$). Thus, the SDE (3.20) has a unique solution. In other words, the martingale problem $\varsigma(\sigma(X_0), X|\mathcal{L}_0, B, C, \nu)$ has unique solution.

Thus, conditions (iii) and (vi) in Theorem 3.3 are satisfied, and the rest conditions are verified easily, since limiting process is continuous, these works are left to the reader. ■

Next, we consider the heavy tailed case. When the distribution of X_1 is heavy-tailed, we will discuss the weak convergence of following processes:

$$\sum_{i=2}^{[nt]} f(\sum_{j=1}^{i-1}(X_{n,j} - E(h(X_{n,j})))(X_{n,i} - E(h(X_{n,i}))),$$

where $X_{n,j} = X_j/b_n$ for some $b_n \to \infty$, $f(x), h(x)$ are continuous functions.

This problem is interesting and difficult from the theoretical point of view. If we use the point process method to obtain the weak convergence, the summation functional should be a continuous functional respect to the Skorohod topology, and the limiting process should have a compound Poisson representation. However, it is difficult to prove that the summation functional above is a continuous functional respect to the Skorohod topology. Moreover, the stochastic integral driven by α–stable Lévy process do not have a compound Poisson representation. Thus, the point process method can not be used easily. By means of Theorem 3.3, we have the following theorem.

Theorem 3.5. *(Lin and Wang (2011)) Let $f : \mathbb{R} \to \mathbb{R}$ be a continuous differentiable function such that*

$$|f(x) - f(y)| \le C|x - y|^a \tag{3.36}$$

for some constants $C > 0$, $a > 0$ and all $x, y \in \mathbb{R}$. Suppose that $\{X_n\}_{n\ge1}$ is a sequence of i.i.d. random variables. Set

$$X_{n,j} = \frac{X_j}{b_n} - \mathbb{E}(h(\frac{X_j}{b_n})) \tag{3.37}$$

for some $b_n \to \infty$, where $h(x)$ is a continuous function satisfying $h(x) = x$ in a neighborhood of 0 and $|h(x)| \le |x|1_{|x|\le1}$. Define measure ρ by

$$\rho((x, +\infty]) = px^{-\alpha}, \qquad \rho([-\infty, -x)) = qx^{-\alpha} \tag{3.38}$$

for $x > 0$, where $\alpha \in (0, 1)$, $0 < p < 1$ and $p + q = 1$. Then

$$(\sum_{i=1}^{[n\cdot]} X_{n,i}, \sum_{i=2}^{[n\cdot]} f(\sum_{j=1}^{i-1} X_{n,j})X_{n,i}) \Rightarrow (Z_\alpha(\cdot), \int_0^{\cdot} f(Z_\alpha(s-))dZ_\alpha(s)) \tag{3.39}$$

in $\mathbb{D}[0, 1]$, where $Z_\alpha(s)$ is an α–stable Lévy process with Lévy measure ρ iff

$$n\mathbb{P}[\frac{X_1}{b_n} \in \cdot] \xrightarrow{v} \rho(\cdot) \tag{3.40}$$

in $\mathbb{M}_p([-\infty, \infty]\backslash\{0\})$.

Set

$$Y_n(t) = \sum_{i=2}^{[nt]} f(\sum_{j=1}^{i-1} X_{n,j})X_{n,i}, \quad Y(t) = \int_0^t f(Z_\alpha(s-))dZ_\alpha(s), \quad S_n(t) = \sum_{i=1}^{[nt]} X_{n,i}.$$

We intend to prove

$$H_n(t) := (Y_n(t), S_n(t)) \Rightarrow H(t) = (Y(t), Z_\alpha(t)).$$

We firstly give some lemmas, which are the basis of the proof.

Lemma 3.6. *The predictable characteristics of $(Y(t), Z_\alpha(t))$ are the triplet (B, C, λ) defined as follows:*

$$\begin{cases} B^i(t) = \begin{cases} \int_0^t \int_{-\infty}^{\infty} (h(f(Z_\alpha(s-)x) - f(Z_\alpha(s-)h(x))\nu(ds, dx), & i = 1, \\ 0, & i = 2, \end{cases} \\ C^{ij}(t) = \begin{cases} \int_0^t \int_{-\infty}^{\infty} h^2(f(Z_\alpha(s-)x)\nu(ds, dx), & i = 1, j = 1, \\ \int_0^t \int_{-\infty}^{+\infty} h(f(Z_\alpha(s-)x)h(x)\nu(ds, dx), & i = 1, j = 2 \text{ or } i = 2, j = 1, \\ \int_0^t \int_{-\infty}^{+\infty} h^2(x)\nu(ds, dx), & i = 2, j = 2, \end{cases} \\ 1_G * \lambda(ds, dx) = 1_G(x, f(Z_\alpha(s-))x)\nu(ds, dx) \text{ for all } G \in \mathbb{B}^2, \end{cases}$$

where $\nu(ds, dx)$ is the compensator of the jump measure of $Z_\alpha(t)$.

Proof. $B^2(t)$ and $C^{22}(t)$ can be easily obtained by the definition of predictable characteristics of a semimartingale.

Let $\eta(ds, dx)$ be the jump random measure of $Y(t)$ and $\lambda'(ds, dx)$ be the compensator of $\eta(ds, dx)$.

If G is a Borel set in $\mathbb{R}$, we have

$$1_G * \lambda'(ds, dx) = 1_G(f(Z_\alpha(s-))x) * \nu(ds, dx).$$

Set $z = f(Z_\alpha(s-))x$, then

$$\begin{aligned} &Y(t) - \int_0^t \int_{-\infty}^{+\infty} (z - h(z))\eta(ds, dz) \\ &= \int_0^t \int_{-\infty}^{+\infty} f(Z_\alpha(s-))h(x)(\mu(ds, dx) - \nu(ds, dx)) \\ &\quad + \int_0^t \int_{-\infty}^{+\infty} f(Z_\alpha(s-))(x - h(x))\mu(ds, dx) \\ &\quad - \int_0^t \int_{-\infty}^{+\infty} (f(Z_\alpha(s-))x - h(f(Z_\alpha(s-)x))\mu(ds, dx) \\ &= \int_0^t \int_{-\infty}^{+\infty} f(Z_\alpha(s-))h(x)(\mu(ds, dx) - \nu(ds, dx)) \\ &\quad + \int_0^t \int_{-\infty}^{+\infty} (h(f(Z_\alpha(s-))x) - f(Z_\alpha(s-)h(x))\mu(ds, dx) \\ &= \int_0^t \int_{-\infty}^{+\infty} f(Z_\alpha(s-))h(x)(\mu(ds, dx) - \nu(ds, dx)) \\ &\quad + \int_0^t \int_{-\infty}^{+\infty} (h(f(Z_\alpha(s-))x) - f(Z_\alpha(s-))h(x))(\mu(ds, dx) - \nu(ds, dx)) \end{aligned}$$

$$+\int_0^t \int_{-\infty}^{+\infty} (h(f(Z_\alpha(s-))x) - f(Z_\alpha(s-))h(x))\nu(ds,dx)$$
$$= \int_0^t \int_{-\infty}^{+\infty} h(f(Z_\alpha(s-))x)(\mu(ds,dx) - \nu(ds,dx))$$
$$+\int_0^t \int_{-\infty}^{+\infty} (h(f(Z_\alpha(s-))x) - f(Z_\alpha(s-))h(x))\nu(ds,dx),$$

which implies

$$B_t^1 = \int_0^t \int_{-\infty}^{+\infty} (h(f(Z_\alpha(s-))x) - f(Z_\alpha(s-))h(x))\nu(ds,dx),$$

and the martingale part of Y_t is

$$\int_0^t \int_{-\infty}^{+\infty} h(f(Z_\alpha(s-))x)(\mu(ds,dx) - \nu(ds,dx)). \tag{3.41}$$

Then we can get C^{11}, C^{12} and C^{21}. The lemma is proved. □

We set

$$\mu_n(\omega; ds, dx) = \sum_{i=1}^n \varepsilon_{(\frac{i}{n}, \frac{X_i(\omega)}{b_n})}(ds, dx),$$

then

$$\nu_n(ds, dx) := \sum_{i=1}^n \varepsilon_{(\frac{i}{n})}(ds)\mathbb{P}(\frac{X_i}{b_n} \in dx)$$

is the compensator of μ_n by independence of $\{X_i\}_{i\geq 1}$. Set

$$\zeta_n(\omega; ds, dx) = \sum_{i=1}^n \varepsilon_{(\frac{i}{n}, \frac{X_i(\omega)}{b_n} - c_n)}(ds, dx),$$

we have

$$\varphi_n(ds, dx) := \sum_{i=1}^n \varepsilon_{(\frac{i}{n})}(ds)\mathbb{P}(\frac{X_i}{b_n} - c_n \in dx)$$

is the compensator of ζ_n, where $c_n = \mathbb{E}[h(\frac{X_1}{b_n})]$.

Firstly, we consider process S_n. Introduce truncate function $h_n(x) = h(x + c_n)$, we have

$$\begin{aligned} S_n(t) &= \sum_{i=1}^{[nt]} h(\frac{X_i}{b_n}) + \sum_{i=1}^{[nt]} (X_{n,i} - h(\frac{X_i}{b_n})) \\ &= \sum_{i=1}^{[nt]} (h(\frac{X_i}{b_n}) - c_n) + \sum_{i=1}^{[nt]} (\frac{X_i}{b_n} - h(\frac{X_i}{b_n})) \\ &= \int_0^t \int_{-\infty}^{+\infty} h(x)(\mu_n(ds,dx) - \nu_n(ds,dx)) + \sum_{i=1}^{[nt]} (\frac{X_i}{b_n} - h(\frac{X_i}{b_n})) \end{aligned}$$

$$=: \widetilde{S}_n(t) + \sum_{i=1}^{[nt]}(\frac{X_i}{b_n} - h(\frac{X_i}{b_n})).$$

The predictable characteristics of $\widetilde{S}_n$ are

$$B_n^2(t) = 0,$$

$$C_n^{22}(t) = \int_0^t \int_{-\infty}^{+\infty} h^2(x)\nu_n(ds, dx) - \sum_{s\leq t}(\int_{-\infty}^{+\infty} h(x)\nu_n(\{s\}, dx))^2.$$

For $Y_n(t)$, we have

$$\begin{aligned}
Y_n(t) &= \sum_{i=2}^{[nt]} h(f(\sum_{j=1}^{i-1} X_{n,j})\frac{X_i}{b_n}) + \sum_{i=2}^{[nt]}(f(\sum_{j=1}^{i-1} X_{n,j})X_{n,i} - h(f(\sum_{j=1}^{i-1} X_{n,j})\frac{X_i}{b_n})) \\
&= \sum_{i=2}^{[nt]}(h(f(\sum_{j=1}^{i-1} X_{n,j})\frac{X_i}{b_n}) - \mathbb{E}(h(f(\sum_{j=1}^{i-1} X_{n,j})\frac{X_i}{b_n})|\mathcal{F}_i)) \\
&\quad + \sum_{i=2}^{[nt]}(\mathbb{E}(h(f(\sum_{j=1}^{i-1} X_{n,j})\frac{X_i}{b_n})|\mathcal{F}_i) - f(\sum_{j=1}^{i-1} X_{n,j})\mathbb{E}(h(\frac{X_1}{b_n}))) \\
&\quad + \sum_{i=2}^{[nt]}(f(\sum_{j=1}^{i-1} X_{n,j})\frac{X_i}{b_n} - h(f(\sum_{j=1}^{i-1} X_{n,j})\frac{X_i}{b_n})) \\
&= \int_0^t \int_{-\infty}^{+\infty} h(f(\sum_{j=1}^{[ns]-1} X_{n,j})x)(\mu_n(ds, dx) - \nu_n(ds, dx)) \\
&\quad + \int_0^t \int_{-\infty}^{+\infty} (h(f(\sum_{j=1}^{[ns]-1} X_{n,j})x) - f(\sum_{j=1}^{[ns]-1} X_{n,j})h(x))\nu_n(ds, dx) \\
&\quad + \sum_{i=2}^{[nt]}(f(\sum_{j=1}^{i-1} X_{n,j})\frac{X_i}{b_n} - h(f(\sum_{j=1}^{i-1} X_{n,j})\frac{X_i}{b_n})) \\
&=: \widetilde{Y}_n(t) + \sum_{i=2}^{[nt]}(f(\sum_{j=1}^{i-1} X_{n,j})\frac{X_i}{b_n} - h(f(\sum_{j=1}^{i-1} X_{n,j})\frac{X_i}{b_n})).
\end{aligned}$$

The predictable characteristics of $\widetilde{Y}_n(t)$ are

$$B_n^1(t) = \int_0^t \int_{-\infty}^{+\infty} (h(f(\sum_{j=1}^{[ns]-1} X_{n,j})x) - f(\sum_{j=1}^{[ns]-1} X_{n,j})h(x))\nu_n(ds, dx),$$

$$C_n^{11}(t) = \int_0^t \int_{-\infty}^{+\infty} h^2(f(\sum_{j=1}^{[ns]-1} X_{n,j})x)\nu_n(ds, dx)$$

$$-\sum_{s\leq t}(\int_{-\infty}^{+\infty} h(f(\sum_{j=1}^{[ns]-1} X_{n,j})x)\nu_n(\{s\},dx))^2,$$

$$C_n^{12}(t) = C_n^{21}(t)$$
$$= \int_0^t \int_{-\infty}^{+\infty} h(f(\sum_{j=1}^{[ns]-1} X_{n,j})x)h(x)\nu_n(ds,dx)$$
$$-\sum_{s\leq t}\int_{-\infty}^{+\infty} h(f(\sum_{j=1}^{[ns]-1} X_{n,j})x)\nu_n(\{s\},dx)\int_{-\infty}^{+\infty} h(x)\nu_n(\{s\},dx).$$

Lemma 3.7. *Under (3.40),*

$$\int_{-\infty}^{+\infty} g(x)nF_n(dx) \to \int_{-\infty}^{+\infty} g(x)\rho(dx), \qquad n\to\infty,$$

for every continuous $g \in \mathbb{C}_2^b(R)$, *where* $F_n(x) = \mathbb{P}(\frac{X_1}{b_n} \leq x)$.

Proof. From (3.40), we have

$$\int_{-\infty}^{+\infty} h(x)nF_n(dx) \to \int_{-\infty}^{+\infty} h(x)\rho(dx), \qquad n\to\infty,$$

for every continuous function h with a compact support.

From (3.38), we can get that for any $\varepsilon > 0$, there exists $r > 0$ such that $\rho((r,+\infty)) + \rho((-\infty,-r)) < \varepsilon$.

Set $B_r = [-r,r]$, we can find a continuous function g_r with a compact support, such that $1_{B_r} \leq g_r \leq 1$. Then

$$|\int_{-\infty}^{+\infty} g(x)nF_n(dx) - \int_{-\infty}^{+\infty} g(x)\rho(dx)|$$
$$\leq |\int_{-\infty}^{+\infty} g(x)nF_n(dx) - \int_{-\infty}^{+\infty} g(x)g_r(x)nF_n(dx)|$$
$$+|\int_{-\infty}^{+\infty} g(x)g_r(x)nF_n(dx) - \int_{-\infty}^{+\infty} g(x)g_r(x)\rho(dx)|$$
$$+|\int_{-\infty}^{+\infty} g(x)g_r(x)\rho(dx) - \int_{-\infty}^{+\infty} g(x)\rho(dx)|$$
$$\leq |\int_{-\infty}^{+\infty} g(x)g_r(x)nF_n(dx) - \int_{-\infty}^{+\infty} g(x)g_r(x)\rho(dx)|$$
$$+||g||(nF_n(B_r^c) - \rho(B_r^c)).$$

For $\varepsilon > 0$, there exists n_0, such that as $n \geq n_0$,

$$|\int_{-\infty}^{+\infty} g(x)g_r(x)nF_n(dx) - \int_{-\infty}^{+\infty} g(x)g_r(x)\rho(dx)| < \varepsilon.$$

From Theorem 3.2 (ii) in Resnick (2007), there exists n_1, such that as $n \geq n_1$,

$$|nF_n(B_r^c) - \rho(B_r^c)| < \varepsilon.$$

Then we have

$$|\int_{-\infty}^{+\infty} g(x)nF_n(dx) - \int_{-\infty}^{+\infty} g(x)\rho(dx)| \leq (1 + 2||g||)\varepsilon$$

as $n \geq \max\{n_0, n_1\}$, which implies the lemma. □

From (3.40), we can obtain

$$\sum_{i=1}^{[n\cdot]} X_{n,i} \Rightarrow Z_\alpha(\cdot)$$

by Theorem 2.7. So $\sum_{i=1}^{[n\cdot]} X_{n,i}$ is relatively compact, in the other words, $\sum_{i=1}^{[n\cdot]} X_{n,i}$ is tight. By this fact, we have that for any $\varepsilon > 0$, there are $n_0 > 0$ and $M > 0$ such that

$$\mathbb{P}(\sup_{t\leq 1} |S_n(t)| > M) < \varepsilon \quad \text{as} \quad n \geq n_0.$$

Since the convergence $H_n \Rightarrow H$ is a local property, it suffices to prove Theorem 3.5 for $f(S_n(t-))1_{[0,T]}$ and $f(Z_\alpha(t-))1_{[0,T]}$ for any stopping time T.

We use S_n^C and S^C to replace T in $f(S_n(t-))1_{[0,T]}$ and $f(Z_\alpha(t-))1_{[0,T]}$ respectively, where $S_n^C = \inf(s : |S_n(s)| \geq C$ or $|S_n(s-)| \geq C)$, $S^C = \inf(s : |S(s)| \geq C$ or $|S(s-)| \geq C)$. As described in Pagès (1986), we can assume

$$f(S_n(t-)) \leq C, \quad f(Z_\alpha(t-)) \leq C \tag{3.42}$$

for some constant C in the following proof.

Let $\mathcal{K}$ be a compact subset of $\mathbb{R}$ such that $|u| \leq C$ for any $u \in \mathcal{K}$. Set

$$1_G * \lambda_n(ds, dx) = 1_G(x, f(\sum_{i=1}^{[ns]-1} X_{n,i})x)\nu_n(ds, dx) \text{ for } G \in \mathbb{B}^2.$$

Lemma 3.8. *Under (3.40), we have that for $t > 0$,*

$$\mathcal{TV}[K * \lambda_n - (K * \lambda) \circ H_n]_t \xrightarrow{\mathbb{P}} 0 \tag{3.43}$$

for every bounded continuous function $K(x, u)$ on $\mathbb{R} \times \mathcal{K}$ satisfying $K(x, u) = 0$ for all $|x| \leq \delta$, $u \in \mathcal{K}$ for some $\delta > 0$.

Proof. At first, we show that for every bounded continuous function g on $\mathbb{R}$,

$$\mathcal{TV}[g * \nu_n - g * \nu]_t \xrightarrow{\mathbb{P}} 0 \quad \text{for } t > 0. \tag{3.44}$$

In fact,

$$\int_0^t \int_{-\infty}^{+\infty} g(x)\nu_n(ds, dx) = [nt]\mathbb{E}(g(X_{n,1}))$$

and

$$\int_0^t \int_{-\infty}^{+\infty} g(x)\nu(ds,dx) = t\int_{-\infty}^{+\infty} g(x)\rho(dx).$$

Hence, we have

$$\mathcal{TV}[g*\nu_n - g*\nu]_t \le \frac{[nt]}{n}\left|\int_{-\infty}^{+\infty} g(x)nF_n(dx) - \int_{-\infty}^{+\infty} g(x)\rho(dx)\right| + \left|\frac{[nt]}{n} - t\right|\int_{-\infty}^{+\infty} g(x)\rho(dx),$$

(3.44) is obtained by Lemma 3.7.

As proved in the Lemma IX5.22 in Jacod and Shiryaev (2003), we only need to prove (3.43) for $K(x,u) = g_a(x)d(x)R(u)$, where $R(u)$ is a continuous function on $\mathcal{K}$, d is a bounded continuous function on $\mathbb{R}$.

Noting that

$$\begin{aligned}
&\mathcal{TV}[K*\lambda_n - (K*\lambda)\circ H_n]_t \\
&\le |R(f(\sum_{i=1}^{[nt]-1} X_{n,i}))|\mathcal{TV}[dg_a*\nu_n - dg_a*\nu]_t + |R(f(S_n(t-))) \\
&\quad - R(f(\sum_{i=1}^{[nt]-1} X_{n,i}))|\cdot(dg_a*\nu)_t \\
&\le ||R||\mathcal{TV}[dg_a*\nu_n - dg_a*\nu]_t + ||d|||R(f(S_n(t-))) - R(f(\sum_{i=1}^{[nt]-1} X_{n,i}))|\cdot(g_a*\nu)_t.
\end{aligned}$$

We can get

$$||R||\mathcal{TV}[dg_a*\nu_n - dg_a*\nu]_t \xrightarrow{\mathbb{P}} 0$$

by (3.44). For any $\varepsilon > 0$, there exists $\delta_1 > 0$, such that $|y - y'| < \delta_1 \Rightarrow |R(y) - R(y')| < \varepsilon$, since $R(u)$ is uniformly continuous on $\mathcal{K}$. Then we have

$$\begin{aligned}
&\mathbb{P}(||d|||R(f(S_n(t-))) - R(f(\sum_{i=1}^{[nt]-1} X_{n,i}))| > \varepsilon) \\
&\le \mathbb{P}(|f(S_n(t-)) - f(\sum_{i=1}^{[nt]-1} X_{n,i})| > \frac{\delta_1}{||d||}) \\
&\le \mathbb{P}(|X_{n,1}| > \frac{\delta_1}{||d||}) \\
&\le 2\frac{\rho(\frac{\delta_1}{||d||},\infty]}{n} \to 0
\end{aligned}$$

by the Lipschitz condition of f and (3.40). Then

$$||d|||R(f(S_n(t-))) - R(f(\sum_{i=1}^{[nt]-1} X_{n,i}))| \xrightarrow{\mathbb{P}} 0, \tag{3.45}$$

which implies

$$||d|||R(f(S_n(t-))) - R(f(\sum_{i=1}^{[nt]-1} X_{n,i}))| \cdot (g_a * \nu)_t \xrightarrow{\mathbb{P}} 0$$

since $g_a * \nu$ is an increasing deterministic measure.

We complete the proof of the lemma. □

Lemma 3.9. *Under (3.40), we have*

$$\mathcal{TV}[B_n^1 - B^1 \circ S_n]_t \xrightarrow{\mathbb{P}} 0 \quad \textit{for any } t > 0.$$

Proof. Let

$$K(x, u) = h(ux) - uh(x).$$

We obtain the lemma by Lemma 3.8. □

Lemma 3.10. *Under (3.40), we have*

$$\mathcal{TV}[C_n^{ij} - C^{ij} \circ S_n]_t \xrightarrow{\mathbb{P}} 0 \quad \textit{for any } t > 0,$$

where $i, j = 1, 2$.

Proof. We only prove the case of $i = j = 1$. The other cases can be proved similarly.

Although this lemma is different from Lemma 3.8, the method of proof is same as that of Lemma 3.8 by noting

$$\mathcal{TV}[h^2(f((\sum_{i=1}^{[nt]-1} X_{n,i})x)) * \nu_n(ds, dx) - h^2(f(Z_\alpha(s-)x)) * \nu(ds, dx) \circ S_n]_t$$

$$\leq |f^2(\sum_{i=1}^{[nt]-1} X_{n,i})|Var[x^2 * \nu_n - x^2 * \nu]_t + |f^2(S_n(t-)) - f^2(\sum_{i=1}^{[nt]-1} X_{n,i})| \cdot (x^2 * \nu)_t$$

$$\leq CVar[x^2 * \nu_n - x^2 * \nu]_t + 2C|f(S_n(t-)) - f(\sum_{i=1}^{[nt]-1} X_{n,i})| \cdot (x^2 * \nu)_t$$

$$\xrightarrow{\mathbb{P}} 0$$

by $|h(x)| \leq |x|1_{|x|\leq 1}$ and Lemma 3.8.

It suffices to show

$$\mathcal{TV}[\sum_{s\leq t}(\int_{-\infty}^{+\infty} h(f(\sum_{j=1}^{[ns]-1} X_{n,j})x)\nu_n(\{s\}, dx))^2] \xrightarrow{\mathbb{P}} 0 \tag{3.46}$$

which is equivalent to

$$\mathcal{TV}[\sum_{s\leq t}(\int_{-\infty}^{+\infty} h(f(\sum_{j=1}^{[ns]-1} X_{n,j})x)\nu_n(\{s\}, dx))^2$$

$$-\sum_{s\le t}(\int_{-\infty}^{+\infty} h(f(Z_\alpha(s-))x)\nu(\{s\},dx))^2 \circ S_n]_t \xrightarrow{\mathbb{P}} 0 \tag{3.47}$$

since $\nu(\{s\},dx))=0$.

However,

$$\mathcal{TV}[h(f(\sum_{j=1}^{[ns]-1} X_{n,j})x) * \nu_n(ds,dx) - h(f(Z_\alpha(s-)x) * \nu(ds,dx) \circ S_n]_t \xrightarrow{\mathbb{P}} 0 \tag{3.48}$$

implies (3.47), and the proof of (3.48) is similar to the argument in the proof of Lemma 3.8. We complete the proof. □

Set

$$1_G * \omega_n(ds,dx) = 1_G(x, f(\sum_{j=1}^{[ns]-1} X_{n,j})x)\varphi_n(ds,dx) \text{ for } G \in \mathbb{B}^2.$$

Lemma 3.11. *Under (3.40), we have that for $t>0$,*

$$\mathcal{TV}[K * \omega_n - (K * \lambda) \circ S_n]_t \xrightarrow{\mathbb{P}} 0$$

for every bounded continuous function $K(x,u)$ on $\mathbb{R}\times\mathcal{K}$ satisfying $K(x,u)=0$ for all $|x|\le\delta$, $u\in\mathcal{K}$ for some $\delta>0$.

Proof. Note that

$$|c_n| \le \mathbb{E}|\frac{X_1}{b_n}|1_{|X_1|\le b_n} = \int_0^1 (\mathbb{P}(|\frac{X_1}{b_n}|>y) - \mathbb{P}(|\frac{X_1}{b_n}|>1))dy \to 0.$$

For $a\neq 0$,

$$n(\mathbb{P}(\frac{X_1}{b_n} - c_n < a) - \mathbb{P}(\frac{X_1}{b_n} < a)) \le n\mathbb{P}(a-|c_n| \le \frac{X_1}{b_n} \le a+|c_n|) \to 0,$$

which implies

$$n\mathbb{P}[\frac{X_1}{b_n} - c_n \in \cdot] \xrightarrow{v} \rho(\cdot) \tag{3.49}$$

by (3.40). From (3.49) and Lemma 3.7, we complete the proof. □

The proof of Theorem 3.5. Assume (3.39) with $f(x)=x$ holds. From Theorem 2.6, we can get (3.40).

Assume that (3.40) holds. we prove (3.39). The proof will be presented in two steps.

(a) We prove the tightness of $\{H_n\}$.

The functions $\alpha \rightsquigarrow B_t(\alpha), C_t(\alpha), g*\lambda_t(\alpha)$ are Skorokhod-continuous on $\mathbb{D}([0,1])$ since the truncation function is continuous. Then $B_n(t), C_n(t), g*\omega_n(t)$ are C-tight by lemmas 3.9-3.11.

From (3.40),

$$\mathcal{L}(S_n) \Rightarrow \mathcal{L}(Z_\alpha).$$

It means that $\{S_n\}$ is tight. Note that

$$\sum_{i=1}^{[nt]} \frac{X_i}{b_n} = S_n(t) + [nt]c_n,$$

and $[nt]c_n \to \int_0^t \int_{-\infty}^{+\infty} h(x)\nu(ds, dx)$. Hence $\{\sum_{i=1}^{[n\cdot]} \frac{X_i}{b_n}\}$ is tight and

$$\lim_{b\uparrow\infty} \limsup_{n\to\infty} P(|x^2|1_{\{|x|>b\}} * \varphi_n(t \wedge S_n^a) > \varepsilon) = 0 \tag{3.50}$$

for all $t > 0$, $a > 0$, $\varepsilon > 0$.

We have

$$\lim_{b\uparrow\infty} \limsup_{n\to\infty} P(|x^2|1_{\{|x|>b\}} * \omega_n(t \wedge S_n^a) > \varepsilon) = 0$$

by (3.42), and hence $\{H_n\}$ is tight.

(b) Identify the limiting process. We need to prove that if for any subsequence weakly converges to a common limit $\widetilde{\mathbb{P}}$, we can identify the limiting process.

Since (3.36), the martingale problem $\varsigma(\sigma(X_0), X|\mathcal{L}_0, B, C, \lambda)$ has unique solution by Theorem 6.13 in *Applebaum (2009)*. So we need to prove the limiting process, H, has predicable characteristics (B, C, λ) under $\widetilde{\mathbb{P}}$, in the other words, to prove

$$h(f(Z_\alpha(s-))x) * (\mu(ds, dx) - \nu(ds, dx)) \circ S_n(t),$$

$$(h(f(Z_\alpha(s-))x) * (\mu(ds, dx) - \nu(ds, dx)) \circ S_n(t))^2 - C^{11} \circ S_n(t),$$

$$g * \eta \circ S_n(t) - g * \lambda \circ S_n(t)$$

are local martingales under $\widetilde{\mathbb{P}}$, where g is a bounded continuous function.

We have

$$\int_0^t \int_{-\infty}^{+\infty} h(f(\sum_{j=1}^{[ns]-1} X_{n,j})x)(\zeta_n(ds, dx) - \varphi_n(ds, dx)) - h(f(Z_\alpha(s-))x)$$
$$*(\mu(ds, dx) - \nu(ds, dx)) \circ S_n(t)$$
$$= h(f((\sum_{i=1}^{[nt]-1} X_{n,i})x) * \varphi_n(ds, dx) - h(f(Z_\alpha(s-)x) * \nu(ds, dx) \circ S_n(t) \xrightarrow{\mathbb{P}} 0 \tag{3.51}$$

by (3.48) and Lemma 3.11.

Set

$$\widetilde{C}_n^{11}(t) = \int_0^t \int_{-\infty}^{+\infty} h^2(f(\sum_{j=1}^{[ns]-1} X_{n,j})x)\varphi_n(ds, dx)$$
$$-\sum_{s\le t}(\int_{-\infty}^{+\infty} h(f(\sum_{j=1}^{[ns]-1} X_{n,j})x)\varphi_n(\{s\}, dx))^2.$$

Since

$$\mathcal{L}(S_n) \Rightarrow \widetilde{\mathbb{P}},$$

(3.42) implies that $C^{11} \circ S_n(t) \leq C$, and lemmas 3.10, 3.11 imply that $\mathbb{P}(\widetilde{C}_n^{11}(1) \geq C+1) \to 0$ as $n \to \infty$. Set $T_n = \inf\{t : \widetilde{C}_n^{11}(t) > C+1\}$, we have

$$\lim_{n\to\infty} \mathbb{P}(T_n < 1) = 0.$$

So

$$\mathbb{E}(\sup_{0\leq t\leq 1} |\int_0^{t\wedge T_n} \int_{-\infty}^{+\infty} h(f(\sum_{j=1}^{[ns]-1} X_{n,j})x)(\zeta_n(ds,dx) - \varphi_n(ds,dx))|^2) \leq 4\mathbb{E}(\widetilde{C}_n^{11}(T_n)) \tag{3.52}$$

by Doob's inequality. (3.51) and (3.52) imply that

$$h(f(Z_\alpha(s-))x) * (\mu(ds,dx) - \nu(ds,dx)) \circ S_n(t)$$

is a local martingale under $\widetilde{\mathbb{P}}$, since

$$\int_0^t \int_{-\infty}^{+\infty} h(f(\sum_{j=1}^{[ns]-1} X_{n,j})x)(\zeta_n(ds,dx) - \varphi_n(ds,dx))$$

is a local martingale.

It is easy to see that

$$\begin{aligned}
&(\int_0^t \int_{-\infty}^{+\infty} h(f(\sum_{j=1}^{[ns]-1} X_{n,j})x)(\zeta_n(ds,dx) - \varphi_n(ds,dx)))^2 \\
&-(h(f(Z_\alpha(s-))x) * (\mu(ds,dx) - \nu(ds,dx)) \circ S_n(t))^2 + C^{11} \circ S_n(t) - \widetilde{C}_n^{11}(t) \\
=& (\int_0^t \int_{-\infty}^{+\infty} h(f(\sum_{j=1}^{[ns]-1} X_{n,j})x)(\zeta_n(ds,dx) - \varphi_n(ds,dx)) \\
&\cdot(\int_0^t \int_{-\infty}^{+\infty} h(f(\sum_{j=1}^{[ns]-1} X_{n,j})x)(\zeta_n(ds,dx) - \varphi_n(ds,dx)) - h(f(Z_\alpha(s-))x) \\
&*(\mu(ds,dx) - \nu(ds,dx)) \circ S_n(t)) \\
&+(h(f(Z_\alpha(s-))x) * (\mu(ds,dx) - \nu(ds,dx)) \circ S_n(t)) \\
&\cdot(\int_0^t \int_{-\infty}^{+\infty} h(f(\sum_{j=1}^{[ns]-1} X_{n,j})x)(\zeta_n(ds,dx) - \varphi_n(ds,dx)) - h(f(Z_\alpha(s-))x) \\
&*(\mu(ds,dx) - \nu(ds,dx)) \circ S_n(t)) \\
&+C^{11} \circ S_n(t) - \widetilde{C}_n^{11}(t).
\end{aligned}$$

Moreover

$$\int_0^{t\wedge T_n} \int_{-\infty}^{+\infty} h(f(\sum_{j=1}^{[ns]-1} X_{n,j})x)(\zeta_n(ds,dx) - \varphi_n(ds,dx))$$

is uniformly integrable by (3.52), thus

$$(\int_0^t \int_{-\infty}^{+\infty} h(f(\sum_{j=1}^{[ns]-1} X_{n,j})x)(\zeta_n(ds,dx) - \varphi_n(ds,dx)))^2$$

$$-(h(f(Z_\alpha(s-))x) * (\mu(ds,dx) - \nu(ds,dx)) \circ S_n(t))^2 + C^{11} \circ S_n(t) - \widetilde{C}_n^{11}(t) \xrightarrow{\mathbb{P}} 0$$

by (3.51) and Lemma 3.10.

By Proposition 3.3,

$$\mathbb{E}(\sup_{0\le t\le 1} |\int_0^{t\wedge T_n}\int_{-\infty}^{+\infty} h(f(\sum_{j=1}^{[ns]-1} X_{n,j})x)(\zeta_n(ds,dx) - \varphi_n(ds,dx))|^4)$$
$$\le C[\mathbb{E}(\widetilde{C}_n^{11}(T_n))^2]^{\frac{1}{2}} + C\mathbb{E}(\widetilde{C}_n^{11}(T_n))^2$$

where C are constants.

$$(h(f(Z_\alpha(s-))x) * (\mu(ds,dx) - \nu(ds,dx)) \circ S_n(t))^2 - C^{11} \circ S_n(t)$$

is local martingale under $\widetilde{\mathbb{P}}$, since

$$(\int_0^t \int_{-\infty}^{+\infty} h(f(\sum_{j=1}^{[ns]-1} X_{n,j})x)(\zeta_n(ds,dx) - \varphi_n(ds,dx)))^2 - \widetilde{C}_n^{11}(t)$$

is a local martingale.

For

$$g * \eta \circ S_n(t) - g * \lambda \circ S_n(t)$$

we can get the similar conclusion by Lemma 3.11. We complete the proof of the theorem.

3.4 Weak convergence to stochastic integral II: Kurtz and Protter's approach

Weak convergence to stochastic integral is a key step in the study of error distribution for approximating a stochastic differential equation. However, the method in previous section may be not suitable when we intend to obtain that convergence of the integrand and integrator implies convergence of the integral under some suitable conditions. *Jakubowski, Mémin and Pagès (1989), Kurtz and Protter (1991)* introduced different approaches to complete this work. Their methods are similar. In this section, we will introduce *Kurtz and Protter's approach* and its applications.

Kurtz and Protter's approach is quite simple, when it deals with the continuous path processes. In this section, we only discuss the continuous path processes. For the processes with jump, the core of method is similar, but the proof is more complex.

First recall that, for every $\delta > 0$, any semimartingale X can be written as

$$X_t = X_0 + A(\delta)_t + M(\delta)_t + \sum_{s\leqslant t} \triangle X_s 1_{\{|\triangle X_s| > \delta\}}, \tag{3.53}$$

where $A(\delta)$ is a predictable process with finite variation, null at 0, $M(\delta)$ is a local martingale null at 0, and $\triangle X_s$ denotes the jump size of X at time s.

Let $\mathbf{X}^n = \{X^n\}$ be a sequence of $\mathbb{R}^d$-valued semimartingales, with $A(\delta)^n$ and $M(\delta)^n$ associated with X^n as in (3.53). We say that the sequence $\{X^n\}$ satisfies $(*)$ if for some $\delta > 0$ and for each i the sequence

$$< M(\delta)^n, M(\delta)^n >_1 + \int_0^1 |dA(\delta)_s^n| + \sum_{0<s\leq 1} |\triangle X_s^n| 1_{\{|\triangle X_s^n > \delta\}}$$

is tight. It turns out that this property is equivalent to the notion of uniform tightness (UT) as introduced by Jakubowski, Mémin and Pagès (1989). Recall that, for any (possibly multidimensional) process V:

$$V^* = \sup_{t\in[0,1]} ||V_t||.$$

The following theorem is the basic tool in this section.

Theorem 3.6. *(Kurtz and Protter (1991)) Let $\{X^n\}$ and $\{Y^n\}$ be two sequences of $\mathbb{R}^d$-valued semimartingales, relative to the filtrations $(\mathcal{F}_t^n)$.*
(a) If both sequences $\{X^n\}$ and $\{Y^n\}$ have $(*)$, then so has the sequence $X^n + Y^n$.
(b) If each X^n is of finite variation and if the sequence $\{\int_0^1 |dX_s^n|\}$ is tight, then the sequence $\{X^n\}$ has $(*)$.
(c) Let $\{H^n\}$ be a sequence of $(\mathcal{F}_t^n)$-predictable processes such that the sequence $\{H^{n*}\}$ is tight. If the sequence $\{X^n\}$ has $(*)$, so has the sequence $\{H^n \cdot X^n\}$.
(d) Let $\{H^n\}$ and $\{H'^n\}$ be two sequences of $(\mathcal{F}_t^n)$-predictable processes such that the sequence $\{H^{n*}\}$ is tight and that $(H^n - H'^n)^* \xrightarrow{P} 0$. If the sequence $\{X^n\}$ has $(*)$, then $(H^n \cdot X^n - H'^n \cdot X^n)^* \xrightarrow{\mathbb{P}} 0$.
(e) Suppose that $\{X^n\}$ weakly converges. Then $(*)$ is necessary and sufficient for the following property: For any sequence $\{H^n\}$ of $(\mathcal{F}_t^n)$-adapted, right-continuous and left-hand limited processes such that the sequence $\{(H^n, X^n)\}$ weakly converges to a limit (H, X), then X is a semimartingale with respect to the filtration generated by the process (H, X), and we have

$$(H^n, X^n, H_-^n \cdot X^n) \Rightarrow (H, X, H_- \cdot X).$$

The proof of Theorem 3.6 can be found in Kurtz and Protter (1991).

In the previous section, we prove Theorem 3.4 by means of strong approximation technique. However, for multivariate case, we can not find a unique stopping time to be embeded into every component of multivariate Brownian motion, so we can not obtain the correspondence results. In this section, we try to use Kurtz and Protter's approach to overcome this difficult.

Let $\{\epsilon_i, \eta_i\}_{i\in\mathbb{Z}}$ be i.i.d. random variabels,

$$u_k = \sum_{j=0}^{\infty} \varphi_j \epsilon_{k-j}, \quad x_k = \sum_{j=0}^{\infty} \delta_j \eta_{k-j},$$

where $\{\varphi_j\}_{j\geq 0}$, $\{\delta_j\}_{j\geq 0}$ are two sequences of numbers.

A fractional Brownian motion with Hurst parameter $0 < H < 1$ on $\mathbb{D}[0,1]$ is defined by

$$B_t^H = \frac{1}{A(H)} \int_{-\infty}^{0} [(t-s)^{H-1/2} - (-s)^{H-1/2}] dW_{-s}^* + \int_0^t (t-s)^{H-1/2} dW_s,$$

where

$$A(H) = (\frac{1}{2H} + \int_0^\infty [(1+s)^{H-1/2} - s^{H-1/2}]^2 ds)^{1/2},$$

$W = (W_s)_{s\geq 0}$ is a standard Brownian motion and $W^* = (W_s^*)_{s\geq 0}$ is an independent copy of W.

Theorem 3.7. *Let $f : \mathbb{R} \to \mathbb{R}$ be a twice differentiable function such that $|f(x)| \leq C(1+|x|^\alpha)$, $|f'(x)| \leq C(1+|x|^\alpha)$ for some constants $C > 0$ and $\alpha > 0$ and all $x \in \mathbb{R}$, and $f''(x)$ is locally bounded. Assume*

(i) $\mathbb{E}(\varepsilon_1) = \mathbb{E}(\eta_1) = 0$, $\mathbb{E}(\varepsilon_1^2) = \mathbb{E}(\eta_1^2) = 1$, $\mathbb{E}|\eta_1|^{4\alpha} < \infty$;

(ii) $\sum_{j=0}^\infty j|\varphi_j| < \infty$;

(iii) $\delta_j \sim j^{-d}\rho(j)$, where $1/2 < d < 1$ and $\rho(j)$ is a function slowly varying at ∞;

Put $d_n^2 = \mathbb{E}(\sum_{k=1}^n x_k)^2$. we have

$$\frac{1}{\sqrt{n}} \sum_{i=1}^{[n\cdot]} f(y_{n,i-1}) u_i \Rightarrow \int_0^\cdot f(B_s^{3/2-d}) dW_s \tag{3.54}$$

in $\mathbb{D}[0,1]$, where $y_{n,k} = \frac{1}{d_n}\sum_{j=1}^k x_j$, W is the Brownian motion.

Furthermore, if (iii) is replaced by

(iv) $\sum_{j=0}^\infty |\delta_j| < \infty$ and $\delta := \sum_{j=0}^\infty \delta_j \neq 0$, we have

$$\frac{1}{\sqrt{n}} \sum_{i=1}^{[n\cdot]} f(y_{n,i-1}) u_i \Rightarrow M \int_0^\cdot f'(G_s) ds + \int_0^\cdot f(B_s) dW_s \tag{3.55}$$

in $\mathbb{D}[0,1]$, where $M = \sum_{j=0}^\infty (\varphi_j (\sum_{s=0}^j \delta_s)) \mathbb{E}[\varepsilon_1 \eta_1]$.

Proof. It is well-know that, with $c_d = \frac{1}{(1-d)(3-2d)} \int_0^\infty x^{-d}(x+1)^{-d} dx$,

$$d_n^2 \sim \begin{cases} c_d n^{3-2d} \rho^2(n) & \text{(iii)}, \\ \delta^2 n & \text{(iv)}. \end{cases}$$

Suppose the $\{X_k\}_{k\geq 1}$ is a tight random variable sequence, $g(x)$ is a locally bounded function. It is obvious that

$$g(\max_{1\leq k\leq n} X_k) = O_{\mathbb{P}}(1).$$

Noting that

$$\frac{1}{\sqrt{n}} \sum_{i=1}^{[nt]} f(y_{n,i-1}) u_i$$

$$= \frac{1}{\sqrt{n}} \sum_{i=1}^{[nt]} f(y_{n,i-1}) \sum_{j=0}^{\infty} \varphi_j \epsilon_{i-j}$$

$$= \frac{1}{\sqrt{n}} \sum_{j=0}^{\infty} \varphi_j \sum_{i=1+j}^{[nt]+j} f(y_{n,i-1})\epsilon_{i-j} + \frac{1}{\sqrt{n}} \sum_{j=0}^{\infty} \varphi_j \sum_{i=1}^{j} f(y_{n,i-1})\epsilon_{i-j}$$

$$- \frac{1}{\sqrt{n}} \sum_{j=0}^{\infty} \varphi_j \sum_{i=[nt]+1}^{[nt]+j} f(y_{n,i-1})\epsilon_{i-j}$$

$$=: \frac{1}{\sqrt{n}} \sum_{j=0}^{\infty} \varphi_j \sum_{i=1+j}^{[nt]+j} f(y_{n,i-1})\epsilon_{i-j} + R([nt]).$$

We first prove that for any $\delta > 0$,

$$\limsup_{n\to\infty} \mathbb{P}\{ \sup_{0\le t\le 1} |R([nt])| \ge \delta\} = 0. \tag{3.56}$$

In fact, we have

$$\frac{1}{n}\mathbb{E} \sup_{0\le t\le 1} (\sum_{j=0}^{\infty} \varphi_j \sum_{i=[nt]+1}^{[nt]+j} f(y_{n,i-1})\epsilon_{i-j})^2$$

$$\le \frac{1}{n} \sum_{j=0}^{\infty} |\varphi_j| \sum_{j=0}^{\infty} j|\varphi_j| [\mathbb{E}(\sup_{[nt]+1\le i\le [nt]+j} f^2(y_{n,i}))]^{1/2}$$

$$\le \frac{C}{n} \sum_{j=0}^{\infty} |\varphi_j| \sum_{j=0}^{\infty} j|\varphi_j| [\mathbb{E}(\sup_{[nt]+1\le i\le [nt]+j} (1+|y_{n,i}|^{\alpha})^2)]^{1/2}$$

and

$$\mathbb{E}[|y_{n,k}|^{2\alpha}] = \frac{1}{d_n^{2\alpha}} \mathbb{E}[| \sum_{i=1}^{k} x_i|^{2\alpha}] = \frac{O((\mathbb{E}[|\sum_{i=1}^{k} x_i|^2])^{\alpha})}{d_n^{2\alpha}} = O(1)$$

by *Wang, Lin and Gulati (2003)*, we can easily obtain (3.56).

Furthermore,

$$\frac{1}{\sqrt{n}} \sum_{j=0}^{\infty} \varphi_j \sum_{i=1+j}^{[nt]+j} f(y_{n,i-1})\epsilon_{i-j}$$

$$= \frac{1}{\sqrt{n}} \sum_{j=0}^{\infty} \varphi_j \sum_{i=1}^{[nt]} f(y_{n,i+j-1})\epsilon_i$$

$$= \frac{1}{\sqrt{n}} \sum_{j=0}^{\infty} \varphi_j \sum_{i=1}^{[nt]} [f(y_{n,i+j-1}) - f(y_{n,i-1})]\epsilon_i + \frac{1}{\sqrt{n}} \sum_{j=0}^{\infty} \varphi_j \sum_{i=1}^{[nt]} f(y_{n,i-1})\epsilon_i$$

and

$$\frac{1}{\sqrt{n}} \sum_{j=0}^{\infty} \varphi_j \sum_{i=1}^{[nt]} [f(y_{n,i+j-1}) - f(y_{n,i-1})]\epsilon_i$$

$$
\begin{aligned}
&= \frac{1}{\sqrt{n}} \sum_{j=0}^{\infty} \varphi_j \sum_{i=1}^{[nt]} f'(y_{n,i-1})(y_{n,i+j-1} - y_{n,i-1})\epsilon_i \\
&+ \frac{1}{2\sqrt{n}} \sum_{j=0}^{\infty} \varphi_j \sum_{i=1}^{[nt]} f''(\xi_{n,i-1})(y_{n,i+j-1} - y_{n,i-1})^2 \epsilon_i,
\end{aligned}
$$

where $\xi_{n,i-1}$ is the random variables between $y_{n,i-1}$ and $y_{n,i+j-1}$.

We have

$$
\begin{aligned}
&|\frac{1}{2\sqrt{n}} \sum_{j=0}^{\infty} \varphi_j \sum_{i=1}^{[nt]} f''(\xi_{n,i-1})(y_{n,i+j-1} - y_{n,i-1})^2 \epsilon_i| \\
&= |\frac{1}{2\sqrt{n} d_n^2} \sum_{j=0}^{\infty} \varphi_j \sum_{i=1}^{[nt]} f''(\xi_{n,i-1})(\sum_{k=i}^{i+j-1} x_k)^2 \epsilon_i| \\
&\leq C \frac{1}{\sqrt{n}} \max_{1 \leq h \leq n} |\varepsilon_h| \sup_{1 \leq h < \infty} x_h^2 \frac{n}{d_n^2} \sum_{j=0}^{\infty} j|\varphi_j| \xrightarrow{\mathbb{P}} 0.
\end{aligned}
$$

Note that

$$
\begin{aligned}
&\frac{1}{\sqrt{n}} \sum_{j=0}^{\infty} \varphi_j \sum_{i=1}^{[nt]} f'(y_{n,i-1})(y_{n,i+j-1} - y_{n,i-1})\epsilon_i \\
&= \frac{1}{\sqrt{n} d_n} \sum_{j=0}^{\infty} \varphi_j \sum_{i=1}^{[nt]} f'(y_{n,i-1})(\sum_{k=i}^{i+j-1} \sum_{s=0}^{\infty} \delta_s \eta_{k-s})\epsilon_i,
\end{aligned}
$$

$$
\sum_{k=i}^{i+j-1} \sum_{s=0}^{\infty} \delta_s \eta_{k-s} = (\sum_{s=0}^{j=1} \delta_s)\eta_i + \sum_{k=i+1}^{i+j-1} (\sum_{s=0}^{j+i-k} \delta_s)\eta_k + \sum_{k=1}^{\infty}(\sum_{s=k}^{j+k} \delta_s)\eta_{i-k}
$$

and

$$
\begin{aligned}
&\mathbb{E}|\sum_{i=1}^{[nt]} f'(y_{n,i-1})(\sum_{k=1}^{\infty}(\sum_{s=k}^{j+k} \delta_s)\eta_{i-k})\epsilon_i|^2 \\
&\leq C \sum_{i=1}^{[nt]} \mathbb{E}[(f'(y_{n,i-1}))^2(\sum_{k=1}^{\infty}(\sum_{s=k}^{j+k} \delta_s)\eta_{i-k})^2] \\
&\leq C \sum_{i=1}^{[nt]} (\mathbb{E}[(f'(y_{n,i-1}))^4])^{1/2} (\mathbb{E}[(\sum_{k=1}^{\infty}(\sum_{s=k}^{j+k} \delta_s)\eta_{i-k})^4])^{1/2},
\end{aligned}
$$

and furthermore,

$$
\mathbb{E}[(f'(y_{n,k}))^4] \leq C(1 + \mathbb{E}[|y_{n,k}|^{4\alpha}]),
$$

$$
\mathbb{E}[|y_{n,k}|^{4\alpha}] = \frac{1}{d_n^{4\alpha}} \mathbb{E}[|\sum_{i=1}^{k} x_i|^{4\alpha}] = \frac{O((\mathbb{E}[|\sum_{i=1}^{k} x_i|^2])^{2\alpha})}{d_n^{4\alpha}} = O(1)
$$

by Wang, Lin and Gulati (2003), and

$$\mathbb{E}[(\sum_{k=1}^{\infty}(\sum_{s=k}^{j+k}\delta_s)\eta_{i-k})^4] \leq C(j(\sum_{h=0}^{\infty}\delta_h^4)).$$

Then

$$\mathbb{E}|\sum_{i=1}^{n} f'(y_{n,i-1})(\sum_{k=1}^{\infty}(\sum_{s=k}^{j+k}\delta_s)\eta_{i-k})\epsilon_i|^2 \leq C\sqrt{j}n. \tag{3.57}$$

By the Kolmogorov inequality for martingale,

$$\begin{aligned}
&\mathbb{E}\sup_{0\leq t\leq 1}(\frac{1}{\sqrt{n}d_n}\sum_{j=0}^{\infty}\varphi_j\sum_{i=1}^{[nt]} f'(y_{n,i-1})(\sum_{k=1}^{\infty}(\sum_{s=k}^{j+k}\delta_s)\eta_{i-k})\varepsilon_i)^2 \\
&\leq \sum_{j=0}^{\infty}|\varphi_j|\sum_{j=1}^{\infty}|\varphi_j|\frac{1}{nd_n^2}\mathbb{E}|\sum_{i=1}^{n} f'(y_{n,i-1})(\sum_{k=1}^{\infty}(\sum_{s=k}^{j+k}\delta_s)\eta_{i-k})\epsilon_i|^2) \\
&\leq \sum_{j=0}^{\infty}|\varphi_j|\sum_{j=1}^{\infty}j|\varphi_j|\frac{1}{nd_n^2} \to 0.
\end{aligned}$$

By the similar method, we can also obtain that

$$\mathbb{E}\sup_{0\leq t\leq 1}(\frac{1}{\sqrt{n}d_n}\sum_{j=0}^{\infty}\varphi_j\sum_{i=1}^{[nt]} f'(y_{n,i-1})((\sum_{k=i+1}^{i+j-1}(\sum_{s=0}^{j+i-k}\delta_s)\eta_k)\varepsilon_i)^2 \to 0, \tag{3.58}$$

$$\mathbb{E}\sup_{0\leq t\leq 1}(\frac{1}{\sqrt{n}d_n}\sum_{j=0}^{\infty}\varphi_j\sum_{i=1}^{[nt]} f'(y_{n,i-1})((\sum_{s=0}^{j}\delta_s)\eta_i - \mathbb{E}(\sum_{s=0}^{j}\delta_s)\eta_i)\varepsilon_i)^2 \to 0, \tag{3.59}$$

Then, we obtain

$$\begin{aligned}
&\sup_{0\leq t\leq 1}|\frac{1}{\sqrt{n}}\sum_{j=0}^{\infty}\varphi_j\sum_{i=1}^{[nt]} f'(y_{n,i-1})(y_{n,i+j} - y_{n,i-1})\epsilon_i \\
&-\frac{1}{\sqrt{n}d_n}\sum_{j=0}^{\infty}\varphi_j\sum_{i=1}^{[nt]} f'(y_{n,i-1})\mathbb{E}((\sum_{s=0}^{j}\delta_s)\eta_i\epsilon_i))| \xrightarrow{\mathbb{P}} 0.
\end{aligned}$$

Thus we just need to discuss the weak convergence of

$$\frac{1}{\sqrt{n}}\sum_{j=0}^{\infty}\varphi_j\sum_{i=1}^{[nt]} f(y_{n,i-1})\epsilon_i + \frac{1}{\sqrt{n}d_n}\sum_{j=0}^{\infty}\varphi_j\sum_{i=1}^{[nt]} f'(y_{n,i-1})\mathbb{E}((\sum_{s=0}^{j}\delta_s)\eta_i\epsilon_i)).$$

Obviously, we have

$$(\frac{1}{\sqrt{n}}\sum_{i=1}^{[n\cdot]}\varepsilon_i, \frac{1}{\sqrt{n}}\sum_{i=1}^{[n\cdot]}\eta_i) \Rightarrow (B, W),$$

where (B, W) is 2-dimensional Brownian motion, $\mathbb{COV}(B_t, W_t) = t\mathbb{E}(\varepsilon_0\eta_0)$.

By the continuous mapping theorem and Theorem 2 in *Sowell (1990)*,

$$\frac{1}{d_n}\sum_{i=1}^{[n\cdot]} x_i \Rightarrow G,$$

$$(\frac{1}{\sqrt{n}}\sum_{i=1}^{[n\cdot]} u_i, f(\frac{1}{d_n}\sum_{i=1}^{[n\cdot]} x_i)) \Rightarrow (W, f(G)), \tag{3.60}$$

$$\frac{1}{n}\sum_{i=2}^{[n\cdot]} f'(y_{n,i}) \Rightarrow \int_0^{\cdot} f'(G_s)ds \tag{3.61}$$

where

$$G_t = \begin{cases} B_t^{3/2-d} & \text{under (iii)}, \\ W_t & \text{under (iv)}. \end{cases}$$

By (iii) and (3.61), we have

$$\sup_{0\le t\le 1} |\frac{1}{\sqrt{n}d_n}\sum_{j=0}^{\infty}\varphi_j\sum_{i=1}^{[nt]} f'(y_{n,i-1})\mathbb{E}((\sum_{s=0}^{j}\delta_s)\eta_i\epsilon_i))| \xrightarrow{\mathbb{P}} 0.$$

(3.60) implies the C-tightness and finite dimensional convergence of $(\frac{1}{\sqrt{n}}\sum_{i=1}^{[n\cdot]} u_i, f(\frac{1}{d_n}\sum_{i=1}^{[n\cdot]} x_i))$, since $W, f(G)$ are continuous path processes. By Theorem 3.6, we obtain (3.54).

Similarly, we can obtain

$$(\frac{1}{\sqrt{n}}\sum_{i=1}^{[n\cdot]} u_i, f(\frac{1}{d_n}\sum_{i=1}^{[n\cdot]} x_i), f'(\frac{1}{d_n}\sum_{i=1}^{[n\cdot]} x_i)) \Rightarrow (W, f(G), f'(G))$$

By (iii) and (3.61), $d_n^2 = \delta^2 n$, we can obtain

$$\frac{1}{d_n^2}\sum_{i=2}^{[n\cdot]} f'(y_{n,i})\mathbb{E}((\sum_{s=0}^{j}\delta_s)\eta_i\epsilon_i)) \Rightarrow K\int_0^{\cdot} f'(G_s)ds.$$

By Theorem 3.6, we complete the proof. □

3.5 Stable central limit theorem for semimartingales

In the study of limit theorems for semimartingales, we usually need to deal with the mixed normal limits. More precisely, we have that

$$Y_n \xrightarrow{d} VN,$$

where $N \sim N(0,1)$, V is a positive random variables independent of N. Usually, the distribution of V is unknown and thus the weak convergence of Y_n can not be used for statistical purposes, since confidence intervals are unavailable.

Furthermore, if V is the weak limit of a sequence $\{V_n\}$, the weak convergence

$$Y_n \xrightarrow{d} VN$$

does not imply

$$(Y_n, V_n) \Rightarrow (VN, V). \tag{3.62}$$

For this reason we need a stronger mode of convergence of Y_n, which would imply the joint weak convergence in (3.62) for random variable V. *Stable convergence in law* is exactly the right type of convergence to guarantee this property.

Firstly, we introduce the *conditional Gaussian martingale* and *martingale biased conditional Gaussian martingale*, which are the limiting processes in stable convergence usually.

We start with a stochastic basis $(\Omega, \mathcal{F}, (\mathcal{F}_t)_{t\in[0,1]}, \mathbb{P})$. The extension of $(\Omega, \mathcal{F}, (\mathcal{F}_t)_{t\in[0,1]}, \mathbb{P})$ is another filtered probability space $(\widetilde{\Omega}, \widetilde{\mathcal{F}}, (\widetilde{\mathcal{F}}_t)_{t\in[0,1]}, \widetilde{\mathbb{P}})$, which is constructed as follows:

$$\widetilde{\Omega} = \Omega\times\Omega', \quad \widetilde{\mathcal{F}} = \mathcal{F}\otimes\mathcal{F}', \quad \widetilde{\mathcal{F}}_t = \bigcap_{s>t} \mathcal{F}_s\otimes\mathcal{F}'_s, \quad \widetilde{\mathbb{P}}(d\omega, d\omega') = \mathbb{P}(d\omega)\mathbb{Q}_\omega d(\omega'), \tag{3.63}$$

where $(\Omega', \mathcal{F}', (\mathcal{F}'_t)_{t\in[0,1]})$ is an auxiliary space, $\mathbb{Q}_\omega d(\omega')$ is a transition probability from $(\Omega, \mathcal{F})$ into $(\Omega', \mathcal{F}')$.

Let $\mathcal{M}_b$ be the set of all bounded martingales on $(\Omega, \mathcal{F}, (\mathcal{F}_t)_{t\in[0,1]}, \mathbb{P})$.

Definition 3.3. A continuous process X on the extension is called an $\mathcal{F}-$conditional Gaussian martingale if X is a local martingale on the extension, orthogonal to all elements of $\mathcal{M}_b$, and $< X, X >$ is $(\mathcal{F}_t)-$adapted.

Let M be a continuous local martingale, and $\mathcal{M}_b(M^\perp)$ be a class of $\mathcal{M}_b$ which are orthogonal to M.

Definition 3.4. A continuous process X on the extension is called an $M-$biased $\mathcal{F}-$conditional Gaussian martingale if it can be written as

$$X_t = X'_t + \int_0^t u_s dM_s,$$

where X' and u are adapted continuous processes.

Now, we recall some facts about stable convergence. Let $\{X_n\}$ be a sequence of random elements defined on $(\Omega, \mathcal{F}, \mathbb{P})$, which are taking values in Polish space E, and X be an $E-$valued random element on the extension space.

Definition 3.5. We say that $\{X_n\}$ stable converges in law to X, and write $X_n \xrightarrow{s-L} X$, if

$$\mathbb{E}(Yf(X_n)) \to \widetilde{\mathbb{E}}(Yf(X))$$

for all $f : E \to \mathbb{R}$ bounded continuous functions and any bounded variable Y on $(\Omega, \mathcal{F})$.

Jacod (1997) gave the following theorem, which is the basic result in stable convergence.

Theorem 3.8. *Let $\{S^n\}$ be a sequence of continuous semimartingales on the stochastic basis $(\Omega, \mathcal{F}, (\mathcal{F}_t)_{t\in[0,1]}, \mathbb{P})$ with predictable characteristics $\{(B^n, C^n, 0)\}$, where $\mathcal{F}$ is separable. Assume that there are two continuous adapted processes C and D, and a continuous bounded variation function B on $(\Omega, \mathcal{F}, (\mathcal{F}_t)_{t\in[0,1]}, \mathbb{P})$ such that*

$$\sup_{t\in[0,1]} |B_t^n - B_t| \xrightarrow{\mathbb{P}} 0 \tag{3.64}$$

$$C_t^n - C_t \xrightarrow{\mathbb{P}} 0 \text{ for any } t \in [0,1], \tag{3.65}$$

$$< M^n, M > -D_t \xrightarrow{\mathbb{P}} 0 \text{ for any } t \in [0,1], \tag{3.66}$$

$$< S^n, N > \xrightarrow{\mathbb{P}} 0 \text{ for any } t \in [0,1] \text{ and any } N \in \mathcal{M}_b(M^\perp), \tag{3.67}$$

where M^n is the local martingale part of S^n, and M is a given martingale. Then, there is a extension of $(\Omega, \mathcal{F}, (\mathcal{F}_t)_{t\in[0,1]}, \mathbb{P})$ and an $M-$biased continuous $\mathcal{F}-$conditional Gaussian martingale S on this extension with

$$< S, S >= C, \qquad < S, M >= D,$$

such that

$$S^n \overset{s-L}{\to} S + B.$$

Proof. It is enough to prove

$$M^n \overset{s-L}{\to} S \tag{3.68}$$

by (3.64). There is a sequence of bounded variables $\{Y_m\}_{m\geq 1}$ which is dense in $\mathcal{L}^1(\Omega, \mathcal{F}, \mathbb{P})$ since $\mathcal{F}$ is separable. Set $N_t^m = \mathbb{E}[Y_m|\mathcal{F}_t]$, so $N^m \in \mathcal{M}_b$, and we have two important properties (c.f. (4.15) in Jacod (1979)):

(i) Every bounded martingale is the limit in $\mathcal{L}^2$, uniformly in time, of a sequence of sums of stochastic integrals with respect to some N^m's.

(ii) $(\mathcal{F}_t)_{t\in[0,1]}$ is the smallest filtration, such that all N^m's are adapted.

By (3.64) and (3.65), we obtain $\{M^n\}$ is tight. Now, we choose any subsequence, indexed by n', such that $\{(M^{n'}, M, N)\}$ converges in law. We can realize the limiting process as follow: consider the canonical space $(\Omega', \mathcal{F}', (\mathcal{F}'_t)_{t\in[0,1]}) = (\mathbb{C}[0,1], \mathcal{C}[0,1], \mathbf{C}[0,1])$ with the canonical process S, and define the extension as (3.63). Furthermore, there is a probability measure $\widetilde{\mathbb{P}}$ on the extension, whose $\Omega-$marginal is $\mathbb{P}$, such that $\{(M^{n'}, M, N)\}$ converges in law to (S, M, N) under $\widetilde{\mathbb{P}}$.

We obtain that S is an $M-$biased continuous conditional Gaussian martingale by Lemma 3.2 and (3.65)-(3.67). Furthermore, the law of S is determined by the processes M, C, D, and it does not depend on the subsequence $\{n'\}$ above. Noting that, in the procedure of proving, we need to prove SN is a local martingale on the extension for any $N \in \mathcal{M}_b(M^\perp)$, which can be implied by (3.67). Hence

$$\mathbb{E}(f(M^n)N_1^m) \to \widetilde{\mathbb{E}}(f(S)N_1^m)$$

for every bounded continuous function f. Furthermore,

$$\mathbb{E}(f(M^n)Y_m) \to \widetilde{\mathbb{E}}(f(S)Y_m)$$

since $\widetilde{\mathbb{E}}(UN_1^m) = \widetilde{\mathbb{E}}(UY_m)$ for any bounded random variable U.

Finally, any bounded random variable Y is the $\mathcal{L}^1$−limit of some subsequence of Y_m, hence

$$\mathbb{E}(f(M^n)Y) \to \widetilde{\mathbb{E}}(f(S)Y)$$

which means (3.68). □

Assume that a sequence $\{\Delta_n\}$ of constants satisfying $\Delta_n \to 0$ as $n \to \infty$, $\{X_{n,i}\}_{i\geq 1}$ is a triangular array of square integrable random variables on the filtered probability space $(\Omega, \mathcal{F}, (\mathcal{F}_t)_{t\in[0,1]}, \mathbb{P})$, where $\mathcal{F}$ is separable, $X_{n,i}$'s are $\mathcal{F}_{i\Delta_n}$−measurable. Theorem 3.8 implies the following theorem easily. In the study of high frequence statistics for stochastic processes X, $X_{n,i}$ is usually equal to $X_{i\Delta_n} - X_{(i-1)\Delta_n}$.

Theorem 3.9. *Assume that there are two absolutely continuous adapted processes u and v, and a continuous bounded variation function B on $(\Omega, \mathcal{F}, (\mathcal{F}_t)_{t\in[0,1]}, \mathbb{P})$ such that*

$$\sup_{t\in[0,1]} \Big|\sum_{i=1}^{[nt]} \mathbb{E}[X_{n,i}|\mathcal{F}_{(i-1)\Delta_n}] - B_t\Big| \xrightarrow{\mathbb{P}} 0, \tag{3.69}$$

$$\sum_{i=1}^{[nt]} (\mathbb{E}[X_{n,i}^2|\mathcal{F}_{(i-1)\Delta_n}] - \mathbb{E}^2[X_{n,i}|\mathcal{F}_{(i-1)\Delta_n}]) - F_t \xrightarrow{\mathbb{P}} 0 \text{ for any } t \in [0,1], \tag{3.70}$$

$$\sum_{i=1}^{[nt]} \mathbb{E}[X_{n,i}(W_{i\Delta_n} - W_{(i-1)\Delta_n})|\mathcal{F}_{(i-1)\Delta_n}] - G_t \xrightarrow{\mathbb{P}} 0 \text{ for any } t \in [0,1], \tag{3.71}$$

$$\sum_{i=1}^{[nt]} \mathbb{E}[X_{n,i}(N_{i\Delta_n} - N_{(i-1)\Delta_n})|\mathcal{F}_{(i-1)\Delta_n}] \xrightarrow{\mathbb{P}} 0 \text{ for any } t \in [0,1] \text{ and any } N \in \mathcal{M}_b(W^\perp), \tag{3.72}$$

$$\sum_{i=1}^{[nt]} \mathbb{E}[X_{n,i}^2 1_{\{|X_{n,i}|>\varepsilon\}}(N_{i\Delta_n} - N_{(i-1)\Delta_n})|\mathcal{F}_{(i-1)\Delta_n}] \xrightarrow{\mathbb{P}} 0 \text{ for any } t \in [0,1] \text{ and any } \varepsilon > 0, \tag{3.73}$$

where W is standard Brownian motion,

$$F_t = \int_0^t (v_s^2 + w_s^2)ds, \quad G_t = \int_0^t v_s ds.$$

Then, there is a extension of $(\Omega, \mathcal{F}, (\mathcal{F}_t)_{t\in[0,1]}, \mathbb{P})$ *and a Brownian motion* W' *on this extension, and independent of* $\mathcal{F}$*, such that*

$$\sum_{i=1}^{[\cdot/\Delta_n]} X_{n,i} \overset{s-L}{\to} B + \int_0^{\cdot} v_s dW_s + \int_0^{\cdot} u_s dW'_s.$$

This theorem is more suitable than Theorem 3.8 to study high frequency statistics.

3.6 An application to stochastic differential equations

Let us consider the following stochastic differential equation (SDE):

$$X_t = x_0 + \int_0^t f(X_{s-})dY_s, \tag{3.74}$$

where f denotes a continuous function and Y is a $\mathbb{R}-$valued semimartingale on $(\Omega, \mathcal{F}, (\mathcal{F}_t)_{0\le t\le 1}, \mathbb{P})$.

Numerical methods for SDEs are a quite important topic. The method provide a numerical solution of SDE, say $\breve{X}$. The law of $\breve{X}$ may be easy to obtain. We can obtain the law of solution of SDE through the known law of $\breve{X}$ and the weak convergence of $\breve{X} - X$.

3.6.1 *Euler method for stochastic differential equations*

In order to study SDE (3.74), we consider the *Euler continuous approximation* X^n to X given by

$$dX_t^n = f(X_{\varphi_n(t)}^n)dY_t,\ X_0^n = x_0, \tag{3.75}$$

where $\varphi_n(t) = [nt]/n$ if $nt \in \mathbb{N}$ and $\varphi_n(t) = t - 1/n$ if $nt \notin \mathbb{N}$, and the *Euler discontinuous approximation* $\overline{X}^n$ given by

$$\overline{X}_t^n = X_{[nt]/n}^n. \tag{3.76}$$

The corresponding *error processes* are denoted by

$$U_t^n = X_t^n - X_t,\ \overline{U}_t^n = \overline{X}_t^n - X_{[nt]/n} = U_{[nt]/n}^n. \tag{3.77}$$

The aim of this section is to find the asymptotic distribution of U^n and $\overline{U}^n$. Weak convergence of stochastic integrals (Theorem 3.6) will play an important role in this section.

Note that

$$U_t^n = \int_0^t (f(X_{\varphi_n(s)}^n) - f(X_{\varphi_n(s)}))dY_s - \int_0^t (f(X_{s-}) - f(X_{\varphi_n(s)}))dY_s, \tag{3.78}$$

and set

$$W_t^n = \int_0^t (f(X_{s-}) - f(X_{\varphi_n(s)}))dY_s.$$

We obtain

$$U_t^n = \int_0^t (f(X_{\varphi_n(s)} + U_{\varphi_n(s)}^n) - f(X_{\varphi_n(s)}))dY_s - W_t^n \tag{3.79}$$
$$= \int_0^t f'(X_{\varphi_n(s)})U_{s-}^n dY_s - W_t^n + o_{\mathbb{P}}(1).$$

We study the weak convergence of U^n, generally, we are interested in convergence of a sequence of SDE's with the form

$$K_t^n = J_t^n + \int_0^t K_{s-}^n H_s^n dY_s, \tag{3.80}$$

where K^n, J^n, H^n are stochastic processes.

The following theorem provides a basis for the results in this section.

Theorem 3.10. *(Jacod and Protter (1998))*

(a) Tightness of both sequences J^{n*} and H^{n*} implies tightness of the sequence K^{n*}.

(b) Suppose that we have another equation (3.80) with solution K'^n and coefficients J'^n and H'^n. If the sequences J^{n*} and H^{n*} are tight and if $(J^n - J'^n)^* \xrightarrow{\mathbb{P}} 0$ and $(H^n - H'^n)^* \xrightarrow{\mathbb{P}} 0$, then $(K^n - K'^n)^* \xrightarrow{\mathbb{P}} 0$.

(c) Let $V_t^n = \int_0^t H_s^n dY_s$. Suppose that the sequence H^{n*} is tight and that the sequence (J^n, V^n) stably converges to a limit (J, V) defined on some extension of the space. Then V is a semimartingale, and

$$(J^n, V^n, K^n) \xrightarrow{s-L} (J, V, K),$$

where K is the unique solution of

$$K_t = J_t + \int_0^t K_{s-} dV_s. \tag{3.81}$$

Proof. Let Z be a càdlàg process on $[0, 1]$ and T a stopping time with respect to $(\mathcal{F}_t)_{0 \le t \le 1}$, define the process Z^{T-} by

$$Z_t^{T-} = Z_t 1_{[0,T)}(t) + Z_{T-} 1_{[T,1]}(t).$$

For

$$K_t = J_t + \int_0^t K_{s-} H_s dY_s, \tag{3.82}$$

let K' is the solution of another equation (3.82) associated with J' and H', and with the same semimartingale Y.

From the slicing technique of Doléans-Dade (see Theorem 5 in Chapter V of *Protter (2005)*), we know for any semimartingale Y and any $\varepsilon > 0$, there is a stopping time T such that $\mathbb{P}(T < 1) \le \varepsilon$, and that the semimartingale $\overline{Y} := Y^{T-}$ is sliceable, i.e.,

$$\mathbb{E}(\sup_{t \in [0,T]} |\int_0^t H_s d\overline{Y}_s|) \le C_{\overline{Y}} \mathbb{E}(H^*),$$

where constant $C_{\overline{Y}}$ only depends on $\overline{Y}$. At last, if $\overline{Y}$ is sliceable and if we consider (3.82) with $|H| \leq A$ a.s. for some constant A, then

$$\mathbb{E}(K^*) \leq C_{A,\overline{Y}}\mathbb{E}(J^*)$$

for a constant $C_{A,\overline{Y}}$ depending on $A, \overline{Y}$.

Fix positive A, ε and u, v, ω, we set

$$S = \inf(t : |H_t| > A \,\text{or}\, |J_t| > u \,\text{or}\, |H_t - H'_t| > v \,\text{or}\, |J_t - J'_t| > \omega) \wedge T$$

and $\overline{J} = J^{S-}, \overline{J}' = J'^{S-}$, and ith component of $\overline{H}$ is $\overline{H}^i = -A \vee \overline{H}^i \wedge A$, and similarly for $\overline{H}'$. We consider the solutions $\overline{K}$ and $\overline{K}'$ of (3.82), associated with $(\overline{J}, \overline{H}, \overline{Y})$ and $(\overline{J}', \overline{H}', \overline{Y}')$ respectively. Note that

$$K = \overline{K}, \;\; K' = \overline{K}' \text{ on the set } \{S > 1\}. \tag{3.83}$$

Note also that $\overline{K}'' = \overline{K}' - \overline{K}$ is the solution of (3.82) associated with $(\overline{J}'', \overline{H}, \overline{Y})$, where

$$\overline{J}''_t = \overline{J}'_t - \overline{J}_t + \int_0^t (\overline{H}'_s - \overline{H}_s)\overline{K}'_{s-}d\overline{Y}_s.$$

Using the properties of sliceable semimartingales, if $v \leq A$ and $\omega \leq u$,

$$\mathbb{E}(\overline{K}''^*) \leq q\mathbb{E}(\overline{J}''^*), \;\; \mathbb{E}(\overline{J}''^*) \leq \omega + qv\mathbb{E}(\overline{K}'^*),$$
$$\mathbb{E}(\overline{K}'^*) \leq (v + \omega)q \leq (A + u)q,$$

where q depends on A, ε and Y. Thus

$$\mathbb{P}(S \leq 1) \leq \varepsilon + \mathbb{P}(H^* > A) + \mathbb{P}(J^* > u) + \mathbb{P}((H - H')^* > v) + \mathbb{P}((J - J')^* > \omega).$$

Hence

$$\begin{aligned}\mathbb{P}((K - K')^* > \eta) \leq \varepsilon &+ \mathbb{P}(H^* > A) + \mathbb{P}(J^* > u) \\ &+ \mathbb{P}((H - H')^* > v) + \mathbb{P}((J - J')^* > \omega) \\ &+ \frac{uv + \omega}{\eta} q,\end{aligned}$$

which, when $J' = 0$ and $H' = 0$, yields

$$\mathbb{P}(K^{n*} > \eta) \leq \varepsilon + 2\mathbb{P}(H^{n*} \geq A) + 2\mathbb{P}(J^{n*} \geq u) + \frac{u}{\eta}C_{\varepsilon,A,Y}, \tag{3.84}$$

where $C_{\varepsilon,A,Y}$ is a constant depending on ε, A and Y. Then we obtain that $\mathbb{P}(K^{n*} > \eta)$ is smaller than $C\varepsilon$, hence the sequence $\{K^{n*}\}$ is tight. Thus (a) and (b) are proved. The assumptions ensure that the sequence $\{V^n\}$ has $(*)$. Since stable convergence is just weak convergence of (U, J^n, V^n) to (U, J, V) for any random variable U on the original probability space, we complete the proof. □

For any process X we write

$$\Delta_i^n X = X_{i/n} - X_{i-1/n}, \quad X_t^{(n)} = X_t - X_{[nt]/n}. \tag{3.85}$$

For any two semimartingales M and N, we write

$$Z_t^n(M,N) = \int_0^t M_{s-}^{(n)} dN_s. \tag{3.86}$$

The fundamental result on the error distribution of Euler method is presented now.

Theorem 3.11. *(Jacod and Protter (1998))*

Let $Z^n = Z^n(Y,Y)$, where Y is a $\mathbb{R}$−valued semimartingale on $(\Omega, \mathcal{F}, (\mathcal{F}_t)_{0\le t\le 1}, \mathbb{P})$, and let α_n be a sequence of positive numbers. There is equivalence between the following statements:

(a) there exists Z on extension of the space on which Y is defined, and the sequence $\alpha_n Z^n$ has (*) and $(Y, \alpha_n Z^n) \Rightarrow (Y, Z)$;

(b) For any x_0 and any differentiable function f with linear growth, there exists U on extension of the space on which Y is defined, and the sequence $\alpha_n U^n$ has (*) and $(Y, \alpha_n U^n) \Rightarrow (Y, U)$;

Under (a) or (b), we can realize the limits Z and U above on the same extension space, and

$$dU_t = f'(X_{t-})[U_{t-}dY_t - f(X_{t-})dZ_t], \; U_0 = 0, \tag{3.87}$$

and $(Y, \alpha_n Z^n, \alpha_n U^n) \Rightarrow (Y, Z, U)$.

This theorem can be easily obtained by Theorem 3.10. The details can be founded in Jacod and Protter (1998).

Applying Theorem 3.11 to continuous local martingale, we have the following theorem.

Theorem 3.12. *Let Y be a continuous local martingale, such that there exist continuous adapted process c such that*

$$C_t = <Y, Y> = \int_0^t c_s ds$$

and assume

$$\int_0^1 c_s^2 ds < \infty. \tag{3.88}$$

Then the sequence $\sqrt{n} Z^n$, defined in Theorem 3.11, stably converges in law to a process Z given by

$$Z_t = \frac{1}{\sqrt{2}} \int_0^t \sigma_s^2 dW_s,$$

where $\sigma^2 = c$, W is a standard Brownian motion defined on an extension of the space on which Y is defined and independent of Y. Moreover, we also have

$$(Y, \sqrt{n}Z^n) \Rightarrow (Y, Z)$$

and

$$(Y, \sqrt{n}U^n) \Rightarrow (Y, U).$$

Proof. Up to enlarging the space, we can assume that there is a Wiener process W' such that

$$Y_t = \int_0^t \sigma_s dW'_s.$$

By Theorem 3.8, if we prove that for all $t \in (0,1]$,

$$D_t^n \xrightarrow{\mathbb{P}} D_t, \qquad \sqrt{n}\langle Z^n, W'\rangle_t \xrightarrow{\mathbb{P}} 0, \tag{3.89}$$

where

$$D_t^n = n < Z^n, Z^n >_t,$$

$$D_t = \frac{1}{2}\int_0^t c_s^2 ds,$$

then the processes $\sqrt{n}Z^n$ will converge stably in law to the process Z. We complete the proof.

At first, we assume there exists $m \in \mathbb{N}$ such that σ has the form

$$\sigma_s = \sum_{i=1}^m A_{i-1} 1_{(t_{i-1}, t_i]}(s), \tag{3.90}$$

where $0 = t_0 < t_1 < \cdots < t_m = 1$ and where A_i is a bounded $\mathcal{F}_{t_i}$−measurable random variable.

Set $\tau_n(u) = u - [nu]/n$. By the Burkholder-Gundy inequality we have, for some constant K,

$$\mathbb{E}(Y_t^{(n)})^4 \leq K/n^2. \tag{3.91}$$

Recall that $Y_t^n = Y_t - Y_{[nt]/n}$. Since $Y_u^n = A_r W_u^{\prime(n)}$ for $t_r \leq [nu]/n \leq u \leq t_{r+1}$, with $B_r = A_r^2$, and for $t_r \leq [nu]/n \leq u \leq v \leq t_{r+1}$,

$$\begin{aligned}
&\mathbb{E}(Y_u^{(n)} Y_v^{(n)} | \mathcal{F}_{s_r}) = B_r \tau_n(u) 1_{\{[nu]=[nv]\}}, \\
&\mathbb{E}((Y_u^{(n)})^2 (Y_v^{(n)})^2 | \mathcal{F}_{s_r}) \\
= \;& \tau_n(u)\tau_n(v)(B_r)^2 + 2\tau_n(u)^2 (B_r)^2 1_{\{[nu]=[nv]\}}.
\end{aligned}$$

Fix r and t such that $0 < t \leq t_{r+1} - t_r$. We have

$$D_{t_r+t}^n - D_{t_r}^n = nB_r \int_{t_r}^{t_r+t} (Y_u^{(n)})^2 du,$$

$$D_{t_r+t} - D_{t_r} = \frac{1}{2}B_r^2 t,$$

$$\sqrt{n}\langle Z^n, W'\rangle_{t_r+t} - \sqrt{n}\langle Z^n, W'\rangle_{t_r} = \sqrt{n}B_r \int_{t_r}^{t_r+t} Y_u^{(n)} du.$$

Set $s(n) = ([nt_r + 1])/n$, then as $n \to \infty$

$$\mathbb{E}[n \int_{s(n)}^{t_r+t} (Y_u^{(n)})^2 du]^2 \to 0 \text{ and } \mathbb{E}[\sqrt{n} \int_{s(n)}^{t_r+t} Y_u^{(n)} du]^2 \to 0$$

by (3.91). So it remains to prove that

$$\mathbb{E}\alpha_n^2 \to 0,$$

where

$$\alpha_n = n \int_{s(n)}^{t_r+t} (Y_u^{(n)})^2 du - \frac{t}{2}B_r.$$

We have

$$\begin{aligned}\mathbb{E}\alpha_n^2 &= n^2 \int_{[s(n),t_r+t]^2} \mathbb{E}(Y_u^{(n)})^2 (Y_v^{(n)})^2 du dv \\ &+ \frac{t^2}{4}\mathbb{E}B_r^2 - nt \int_{s(n)}^{t_r+t} \mathbb{E}(B_r (Y_u^{(n)})^2) du.\end{aligned}$$

On the one hand,

$$nt \int_{s(n)}^{t_r+t} \mathbb{E}(B_r (Y_u^{(n)})^2) du \to \frac{t^2}{2}\mathbb{E}B_r^2.$$

On the other hand,

$$\left|\mathbb{E}(Y_u^{(n)})^2 (Y_v^{(n)})^2 - \tau_n(u)\tau_n(v)\mathbb{E}B_r^2\right| \le \frac{K}{n^2} 1_{\{[nu]=[nv]\}},$$

and thus

$$n^2 \int_{[s(n),t_r+t]^2} \mathbb{E}(Y_u^{(n)})^2 (Y_v^{(n)})^2 du dv \to \frac{t^2}{4}\mathbb{E}B_r^2.$$

Thus

$$\mathbb{E}\alpha_n^2 \to 0,$$

and we have (3.89) for the case of (3.90).

It remains to prove that (3.88) implies (3.89) in the general case.

Let $T_p = \inf(t : \int_0^t c_s^2 ds \ge p)$. Since (3.88) yields $\mathbb{P}(T_p < 1) \to 0$ as $p \to \infty$ by localization, it is enough to prove the result for the processes stopped at time T_p. From now on, we assume that

$$\int_0^1 c_s^2 ds \le p.$$

There exists a sequence $\sigma(r)$ of processes of the form (3.90), such that

$$\eta_r := \int_0^1 |\sigma_s - \sigma(r)_s|^4 ds \to 0,$$

$$\int_0^1 |\sigma(r)_s|^4 ds \leq \int_0^1 \sigma_s^4 ds \leq p.$$

Let $Y(r)_t = \int_0^t \sigma(r)_s dW'_s$ with the associated processes $Z(r)^n$ and $D(r)^n$ and $D(r)$. In fact, for each r we have that the following converge for all t

$$D(r)_t^n \xrightarrow{\mathbb{P}} D(r)_t, \qquad \sqrt{n}\langle Z(r)^n, W'\rangle_t \xrightarrow{\mathbb{P}} 0.$$

We have, with $c(r) = \sigma(r)^2$,

$$\begin{aligned}
|D(r)_t^n - D_t^n| &= n\left|\int_0^t ((Y(r)_s^{(n)})^2 c(r)_s - (Y_s^{(n)})^2 c_s) ds\right| \\
&\leq n\int_0^t (Y(r)_s^{(n)})^2 |\sigma(r)_s - \sigma_s|(|\sigma(r)_s| + |\sigma_s|) ds \\
&\quad + \int_0^t |Y(r)_s^{(n)} - Y_s^{(n)}|(|Y(r)_s^{(n)}| + |Y_s^{(n)}|)\sigma_s^2 ds.
\end{aligned}$$

By the Burkholder-Gundy inequality and the Cauchy-Schwarz inequality,

$$\mathbb{E}(|Y_s^{(n)}|^4) \leq \frac{K}{n}\mathbb{E}\left(\int_{[ns]/n}^s |\sigma(q)_u|^4 du\right),$$

thus

$$\int_0^t \mathbb{E}(|Y_s^{(n)}|^4) ds \leq \frac{K}{n^2}$$

for some constant K, and also

$$\int_0^t \mathbb{E}(|Y(r)_s^{(n)}|^4) ds \leq \frac{K}{n^2}.$$

The same argument shows that

$$\int_0^t \mathbb{E}(|Y(r)_s^{(n)} - Y_s^{(n)}|^4) ds \leq \frac{K\eta_r}{n^2}.$$

Thus

$$\mathbb{E}(|D(r)_t^n - D_t^n|) \leq K\eta_r^{1/4},$$

which implies the first part of (3.89). The second part of (3.89) is proved similarly. □

Random grid scheme for SDEs is an interesting subject. Comparing to the equidistant determinsitic grid, random grid scheme can be chosen to design the approximation error so that it has desirable properties.

Lindbwerg and Rootzén (2013) considered the random grid scheme, which is based on the following stopping times, $\tau_0^n = 0$,

$$\tau_{k+1}^n = (\tau_k^n + \frac{1}{n\theta(\tau_k^n)}) \wedge 1$$

for some adapted stochastic process θ. Let

$$\eta_n(t) = \tau_k^n, \quad \tau_k^n \le t < \tau_{k+1}^n$$

for $k = 1, 2, \cdots$.

Theorem 3.13. *(Lindbwerg and Rootzén (2013)) Let the measurable functions* $\alpha(\cdot) : \mathbb{R} \to \mathbb{R}$, $\beta(\cdot) : \mathbb{R} \to \mathbb{R}$ *satisfy*

$$|\alpha(x)| + |\beta(x)| \le C(1 + |x|),$$

where $x \in \mathbb{R}$ *for some constant* C *and*

$$|\alpha(x) - \alpha(y)| + |\beta(x) - \beta(y)| \le D(|x - y|),$$

where $x, y \in \mathbb{R}$ *for some constant* D. *Let* Y *be the solution of the SDE*

$$dY(t) = \alpha(Y(t))dt + \beta(Y(t))dW(t), \tag{3.92}$$

where W *is a Brownian motion and* $Y(0)$ *is independent of* W *and satisfies* $\mathbb{E}Y(0)^2 < \infty$. *Furthermore, assume*

$$\sup_{t \in [0,1]} \theta(t) < \infty \quad a.s.$$

and $1/\theta$ *is Riemann integrable. The error in the Euler scheme is defined by*

$$\widetilde{U}_t^n = \sqrt{n} \int_0^t (f(Y(u)) - f(Y(\eta_n(u))))dY(u).$$

where f *is a continuously differentiable function. Then*

$$\sqrt{n}\widetilde{U}^n \Rightarrow \widetilde{U}$$

on $\mathbb{C}[0, 1]$, *where*

$$\widetilde{U}_t = \int_0^t \frac{f'(Y(u))\beta^2(Y(u))}{\sqrt{2\theta(u)}} dB(u)$$

and B *is a Brownian motion independent of* W.

Proof. We first assume α and β are uniformly bounded. There exists a unique solution Y of (3.92) by Theorem 5.2.1 in *Oksendal (2003)*. By Theorem 3.11, we firstly discuss the joint weak convergence of (Z^n, Y), where

$$Z^n(t) = \sqrt{n} \int_0^t (Y(s) - Y(\eta_n(s)))dY(s).$$

We first show that

$$\sqrt{n}\sup_{t\in[0,1]}|\int_0^t\sum_{i=1}^{\infty}\int_{\tau_i^n}^s 1_{\{\tau_i^n\le s<\tau_{i+1}^n\}}\beta(Y(u))dW(u)ds|\xrightarrow{\mathbb{P}}0. \tag{3.93}$$

Write

$$\begin{aligned}
&\sqrt{n}\int_0^t\sum_{i=1}^{\infty}\int_{\tau_i^n}^s 1_{\{\tau_i^n\le s<\tau_{i+1}^n\}}\beta(Y(u))dW(u)ds\\
&=\sqrt{n}\int_0^t\sum_{i=1}^{\infty}\int_{\tau_i^n}^s 1_{\{\tau_i^n\le s<\tau_{i+1}^n\}}(\beta(Y(u))-\beta(Y(\tau_i^n)))dW(u)ds\\
&\quad+\sqrt{n}\int_0^t\sum_{i=1}^{\infty}\int_{\tau_i^n}^s 1_{\{\tau_i^n\le s<\tau_{i+1}^n\}}\beta(Y(\tau_i^n))dW(u)ds\\
&=: I_1+I_2.
\end{aligned}$$

In fact

$$I_2=\int_0^t\sqrt{n}\sum_{i=1}^{\infty}1_{\{\tau_i^n\le s<\tau_{i+1}^n\}}\beta(Y(\tau_i^n))(W(s)-W(\tau_i^n))ds.$$

We have that

$$\begin{aligned}
&\sqrt{n}\mathbb{E}[\int_{\tau_i^n}^{\tau_{i+1}^n}\beta(Y(\tau_i^n))(W(s)-W(\tau_i^n))ds|\mathcal{F}_{\tau_i^n}]\\
&=\sqrt{n}\beta(Y(\tau_i^n))\int_0^{1/n\theta(\tau_i^n)}\mathbb{E}W(s)ds\\
&=0,
\end{aligned}$$

and

$$M_i^n:=\int_0^{\tau_i^n}\sqrt{n}\sum_{k=1}^{\infty}1_{\{\tau_k^n\le s<\tau_{k+1}^n\}}\beta(Y(\tau_k^n))(W(s)-W(\tau_i^n))ds$$

is a martingale. Using the Cauchy-Schwarz inequality,

$$\begin{aligned}
&\sum_{i=1}^{\infty}\mathbb{E}[(M_{i+1}^n-M_i^n)^2|\mathcal{F}_{\tau_i^n}]\\
&=\sum_{i=1}^{\infty}\mathbb{E}[(\int_{\tau_i^n}^{\tau_{i+1}^n}\sqrt{n}\sum_{k=1}^{\infty}1_{\{\tau_k^n\le s<\tau_{k+1}^n\}}\beta(Y(\tau_k^n))(W(s)-W(\tau_i^n))ds)^2|\mathcal{F}_{\tau_i^n}]\\
&\le n\sum_{i=1}^{\infty}\beta^2(Y(\tau_i^n))(\tau_{i+1}^n-\tau_i^n)\int_{\tau_i^n}^{\tau_{i+1}^n}\mathbb{E}[(W(s)-W(\tau_i^n))^2|\mathcal{F}_{\tau_i^n}]ds\\
&=\frac{n}{4}\sum_{i=1}^{\infty}\beta^2(Y(\tau_i^n))(\tau_{i+1}^n-\tau_i^n)^3\\
&\le\frac{1}{4}\max_i(\frac{\beta(Y(\tau_i^n))}{n\theta^{3/2}(\tau_i^n)})\sum_{\tau_i^n<1}(\frac{\beta(Y(\tau_i^n))}{n\theta^{3/2}(\tau_i^n)})\to 0.
\end{aligned}$$

Thus

$$\max_i |M_i^n| \xrightarrow{\mathbb{P}} 0 \tag{3.94}$$

by the Doob inequality of martingale.

Moreover,

$$\max_i \sup_{\tau_i^n \le t < \tau_{i+1}^n} \int_{\tau_i^n}^t \sqrt{n} \sum_{k=1}^{\infty} 1_{\{\tau_k^n \le s < \tau_{k+1}^n\}} \beta(Y(\tau_k^n))(W(s) - W(\tau_i^n)) ds$$

$$\le \max_i \{\tau_{i+1}^n - \tau_i^n\}^{1/2} \Big(\int_0^1 (\sqrt{n} \sum_{k=1}^{\infty} 1_{\{\tau_k^n \le s < \tau_{k+1}^n\}} \beta(Y(\tau_k^n))(W(s) - W(\tau_i^n)))^2 ds \Big)^{1/2}$$

$$\xrightarrow{\mathbb{P}} 0. \tag{3.95}$$

Combining (3.94) with (3.95) implies

$$I_2 \xrightarrow{\mathbb{P}} 0. \tag{3.96}$$

Consider I_1. By Itô's isometry,

$$\mathbb{E}\Big[\Big(\int_{\tau_i^n}^s 1_{\{\tau_i^n \le s < \tau_{i+1}^n\}} (\beta(Y(u)) - \beta(Y(\tau_i^n))) dW(u)\Big)^2 | \mathcal{F}_{\tau_i^n}\Big]$$

$$= \mathbb{E}\Big[\int_{\tau_i^n}^s 1_{\{\tau_i^n \le s < \tau_{i+1}^n\}} (\beta(Y(u)) - \beta(Y(\tau_i^n)))^2 du | \mathcal{F}_{\tau_i^n}\Big]$$

$$\le C \int_{\tau_i^n}^{\tau_{i+1}^n} \mathbb{E}[(Y(u) - Y(\tau_i^n))^2 | \mathcal{F}_{\tau_i^n}] du$$

$$\le C \int_{\tau_i^n}^{\tau_{i+1}^n} (u - \tau_i^n) du \le C(\tau_{i+1}^n - \tau_i^n)^2.$$

Let

$$\Delta_i^n(t) = \sqrt{n} \int_{\tau_i^n}^{\tau_{i+1}^n \wedge t} \int_{\tau_i^n}^{\tau_{i+1}^n \wedge s} (\beta(Y(u)) - \beta(Y(\tau_i^n))) dW(u) ds,$$

Using the Doob inequality and the Cauchy-Schwarz inequality, we have

$$\mathbb{E}\Big[\sup_{\tau_i^n \le t < \tau_{i+1}^n} |\Delta_i^n(t)| | \mathcal{F}_{\tau_i^n}\Big]$$

$$\le \sqrt{n}(\tau_{i+1}^n - \tau_i^n) \mathbb{E}\Big[\sup_{\tau_i^n \le s < \tau_{i+1}^n} \Big| \int_{\tau_i^n}^s (\beta(Y(u)) - \beta(Y(\tau_i^n))) dW_j(u) \Big| | \mathcal{F}_{\tau_i^n}\Big]$$

$$\le C\sqrt{n}(\tau_{i+1}^n - \tau_i^n) \mathbb{E}\Big[\int_{\tau_i^n}^{\tau_{i+1}^n} (\beta(Y(u)) - \beta(Y(\tau_i^n)))^2 du | \mathcal{F}_{\tau_i^n}\Big]^{1/2}$$

$$\le C\sqrt{n}(\tau_{i+1}^n - \tau_i^n)^2.$$

Thus

$$\sum_{i=1}^{\infty} \mathbb{E}\Big[\sup_{\tau_i^n \le t < \tau_{i+1}^n} |\Delta_i^n(t)| | \mathcal{F}_{\tau_i^n}\Big] \le C\sqrt{n} \sum_{i=1}^{\infty} (\tau_{i+1}^n - \tau_i^n)^2$$

$$\leq \frac{C}{\sqrt{n}} \sup_{0\leq t\leq 1} \frac{1}{\theta(t)} \xrightarrow{a.s.} 0.$$

Hence,

$$I_1 \xrightarrow{a.s.} 0.$$

Combining it with (3.96) implies (3.93).

Completely similar, but more complex computation shows that

$$\begin{aligned}
& n \int_0^t \sum_{k=1}^{\infty} \int_{\tau_k^n}^s 1_{\{\tau_k^n \leq s < \tau_{k+1}^n\}} \beta(Y(u)) dW(u) \\
& \quad \cdot \int_{\tau_k^n}^s 1_{\{\tau_k^n \leq s < \tau_{k+1}^n\}} \beta(Y(z)) dW_m(z) \beta^2(Y(s)) ds \\
= {} & n \int_0^t \sum_{k=1}^{\infty} 1_{\{\tau_k^n \leq s < \tau_{k+1}^n\}} \beta^4(Y(\tau_k^n)) \cdot (W(s) - W(\tau_k^n))^2 ds + o_{\mathbb{P}}(1) \\
\xrightarrow{\mathbb{P}} {} & \frac{1}{2} \int_0^t \beta^4(Y(s))/\theta(s) ds.
\end{aligned}$$

By Theorem 3.8 and Theorem 3.11, we obtain the result under the uniformly boundness of α and β.

To remove the restriction, we can apply the local procedure, which is similar to that in Theorem 3.12 and is left to the reader. □

Remark 3.2. Theorem 3.13 can be extend to the multidimentional case by similar argument.

3.6.2 *Milstein method for stochastic differential equations*

The *Milstein scheme* can improve the rate of convergence for Euler method from $1/n$ to $1/n^2$. In this subsection, we study the normalized asymptotic error for Milstein scheme. Let Y be an $\mathbb{R}-$valued continuous semimartingale on $(\Omega, \mathcal{F}, (\mathcal{F}_t)_{0\leq t\leq 1}, \mathbb{P})$ with $Y_0 = 0$. Consider the SDE (3.74), here we assume f is a twice order differentiable function, and satisfies

$$|f(x)| \leq C(1 + |x|)$$

for some constant C.

The Milstein scheme is defined by $\hat{X}_0^n = 0$, and

$$\hat{X}_t^n = \hat{X}_{\varphi_n(t)}^n + f(\hat{X}_{\varphi_n(t)}^n)(Y_t - Y_{\varphi_n(t)}) + \int_{\varphi_n(t)}^t g(\hat{X}_{\varphi_n(t)}^n)(Y_s - Y_{\varphi_n(s)}) dY_s,$$

where $g(x) = f'(x)f(x)$, $\varphi_n(t) = [nt]/n$ if $nt \in \mathbb{N}$ and $\varphi_n(t) = t - 1/n$ if $nt \bar{\in} \mathbb{N}$.

As above, our concern is

$$\hat{U}_t^n := \hat{X}_t^n - X_t.$$

We first introduce some notation below. For any process V, we write $\Delta_t^n(V) = V_t^{(n)} := V_t - V_{\varphi_n(t)}$, and

$$Z_t^n = \int_0^t Y_s^{(n)} dY_s,$$
$$M_t^n = \int_0^t \int_{\varphi_n(s)}^s Y_r^{(n)} dY_r dY_s,$$
$$N_t^n = \int_0^t (Y_s^{(n)})^2 dY_s,$$
$$R_t^n = \int_0^t g(\hat{X}_{\varphi_n(s)}^n) Y_s^{(n)} dY_s.$$

Obviously,

$$\hat{X}_t^n = \hat{X}_{\varphi_n(t)}^n + f(\hat{X}_{\varphi_n(t)}^n)(Y_t - Y_{\varphi_n(t)}) + R_t^n,$$

and therefore,

$$\Delta_t^n(\hat{X}^n) = f(\hat{X}_{\varphi_n(t)}^n)\Delta_t^n(Y) + \Delta_t^n(R^n).$$

Integrating the both sides with respect to $f'(\hat{X}_s^n)dY_s$, then

$$\int_0^t \Delta_s^n(\hat{X}^n) f'(\hat{X}_{\varphi_n(s)}^n) dY_s = \int_0^t \Delta_s^n(R^n) f'(\hat{X}_{\varphi_n(s)}^n) dY_s + R_t^n. \tag{3.97}$$

Thus

$$\hat{U}_t^n = \int_0^t ([f(\hat{X}_s^n) - f(X_s)] - [f(\hat{X}_s^n) - f(\hat{X}_{\varphi_n(s)}^n)]) dY_s + R_t^n.$$

By Taylor's expansion, there exist $\bar{\xi}_s^n$ between $\hat{X}_s^n$ and X_s, and ξ_s^n between $\hat{X}_s^n$ and $\hat{X}_{\varphi_n(s)}^n$ such that

$$f(\hat{X}_s^n) - f(X_s) = f'(\bar{\xi}_s^n)\hat{U}_s^n,$$
$$f(\hat{X}_s^n) - f(\hat{X}_{\varphi_n(s)}^n) = f'(\hat{X}_{\varphi_n(s)}^n)\Delta_s^n(\hat{X}^n) + \frac{1}{2} f''(\xi_s^n)(\Delta_s^n(\hat{X}^n))^2.$$

Therefore,

$$\hat{U}_t^n = \int_0^t f'(\bar{\xi}_s^n)\hat{U}_s^n dY_s + R_t^n$$
$$- \int_0^t f'(\hat{X}_{\varphi_n(s)}^n)\Delta_s^n(\hat{X}^n) dY_s - \frac{1}{2}\int_0^t f''(\xi_s^n)(\Delta_s^n(\hat{X}^n))^2 dY_s.$$

From (3.97),

$$\hat{U}_t^n = \int_0^t f'(\bar{\xi}_s^n)\hat{U}_s^n dY_s$$
$$- \int_0^t f'(\hat{X}_{\varphi_n(t)}^n)\Delta_s^n(R^n) dY_s - \frac{1}{2}\int_0^t f''(\xi_s^n)(\Delta_s^n(\hat{X}^n))^2 dY_s.$$

In fact

$$\int_0^t f'(\hat{X}^n_{\varphi_n(s)})\Delta^n_s(R^n)dY_s = \int_0^t h(\hat{X}^n_{\varphi_n(s)})dM^n_s,$$

where $h(x) = f(x)(f'(x))^2$. Hence

$$\hat{U}^n_t = \int_0^t f'(\bar{\xi}^n_s)\hat{U}^n_s dY_s - \int_0^t h(\hat{X}^n_{\varphi_n(s)})dM^n_s - \frac{1}{2}\int_0^t f''(\xi^n_s)(\Delta^n_s(\hat{X}^n))^2 dY_s.$$

Note that

$$\sup_{0\le t\le 1} |\int_0^t f''(\xi^n_s)(\Delta^n_s(\hat{X}^n))^2 dY_s - \int_0^t f''(\xi^n_s)f^2(\hat{X}^n_{\varphi_n(s)})dN^n_s| \xrightarrow{\mathbb{P}} 0.$$

Through simple computation, we can easily obtain Theorem 3.11 for Milstein method, then the following theorem holds.

Theorem 3.14. *(Yan (2005)) Let α_n be a constants sequence of positive numbers. There is equivalence between the following statements*
(a) The sequence $(\alpha_n M^n, \alpha N^n)$ has $(*)$ and

$$(Y, \alpha_n M^n, \alpha_n N^n) \Rightarrow (Y, M, N);$$

(b) For any x_0 and any twice order differentiable function f with linear growth, the sequence $\alpha_n \hat{U}^n$ has $(*)$ and

$$(Y, \alpha_n \hat{U}^n) \Rightarrow (Y, \hat{U}).$$

In this case, we can realize the limits M, N and $\hat{U}$ above on the same extension of the space on which Y is defined, and

$$d\hat{U}_t = \hat{U}_t f'(X_t)dY_t - f(X_t)(f')^2(X_t)dM_t - f''(X_t)f^2(X_t)dN_t \tag{3.98}$$

and $\hat{U}_0$=0.

3.7 Appendix: the predictable characteristics of semimartingales

Drift, variance of Gaussian part and Lévy measure of Lévy process play an important role in limit theory. The predictable characteristics of semimartingales are introduced to replace these three terms. In this section, we will list the basic concept and property of predictable characteristics of semimartingales. More details and the proofs of propositions can be found in Jacod and Shiryaev (2003).

We first introduce the definition of random measure and its related property.

Definition 3.6. A *random measure* on $\mathbb{R}_+ \times \mathbb{R}$ is a family $\mu = (\mu(\omega; dt, dx) : \omega \in \Omega)$ of nonnegative measures on $(\mathbb{R}_+ \times \mathbb{R}, \mathcal{B}_+ \otimes \mathcal{B})$ satisfying $\mu(\omega; \{0\} \times \mathbb{R}) = 0$ identically.

To introduce the compensator of random measure, we need to put: $\widetilde{\Omega} = \Omega \times \mathbb{R}_+ \times \mathbb{R}$ with the $\sigma-$fields $\widetilde{\mathcal{O}} = \mathcal{O} \times \mathcal{B}$, $\widetilde{\mathcal{P}} = \mathcal{P} \times \mathcal{B}$, where $\mathcal{O}$ is the optional field of $\Omega \times \mathbb{R}_+$, $\mathcal{P}$ is the predictable field of $\Omega \times \mathbb{R}_+$.

Definition 3.7. An optional (predictable) function on $\widetilde{\Omega}$ is a function which are measurable respect to the fields $\widetilde{\mathcal{O}}$ $(\widetilde{\mathcal{P}})$.

H is a optional function on $\widetilde{\Omega}$, for every $\omega \in \Omega$, the integral is defined by

$$H * \mu_t(\omega) = \begin{cases} \int_{[0,t]\times\mathbb{R}} H(\omega, t, x)\mu(\omega; dt, dx) \text{ if } \int_{[0,t]\times\mathbb{R}} |W(\omega, t, x)|\mu(\omega; dt, dx) < \infty; \\ 0, \qquad\qquad\qquad\qquad\qquad\qquad\qquad \text{otherwise.} \end{cases}$$

Definition 3.8. A random measure μ is called optional (predictable) if $W * \mu$ is optional (predictable) for every optional (predictable) function W.

Definition 3.9. An optional random measure μ is called $\widetilde{\mathcal{P}}-\sigma-$finite if there exists a strict positive predictable function V such that $V * \mu_\infty < \infty$.

Now, we introduce the definition of compensator of a random measure.

Definition 3.10. The compensator of optional $\widetilde{\mathcal{P}} - \sigma-$finite random measure μ, denoted by μ^p, is a predictable random measure such that

$$\mathbb{E}(W * \mu_\infty) = \mathbb{E}(W * \mu^p_\infty)$$

for any positive predictable function W.

To characterize the jump of underlying semimartingale, we need to study the integer valued random measure.

Definition 3.11. An integer valued optional $\widetilde{\mathcal{P}} - \sigma-$finite random measure μ with $\mu(\omega, t \times \mathbb{R}) \leq 1$ is called an integer valued random measure.

For a semimartingale X,

$$\mu^X(\omega; dt, dx) := \sum_s 1_{\{\triangle X_s(\omega) \neq 0\}} \varepsilon_{(s, \triangle X_s)}(dt, dx)$$

is obviously an integer valued random measure.

Now, we introduce the definition of predictable characteristics of a semimartingale.

Let $(\Omega, \mathcal{F}, \{\mathcal{F}_t\}, \mathbb{P})$ be a filtered probability space, X be a semimartingale defined on $(\Omega, \mathcal{F}, \{\mathcal{F}_t\}_{t\geq 0}, \mathbb{P})$. Set $h(x) = x1_{|x|\leq 1}$, and

$$\begin{cases} \check{X}(h)_t = \sum_{s\leq t}[\Delta X_s - h(\Delta X_s)], \\ X(h) = X - \check{X}(h). \end{cases} \tag{3.99}$$

$X(h)$ is a special semimartingale, a semimartingale with predictable finite variations part, and we consider its canonical decomposition

$$X(h) = X_0 + M(h) + B(h) \tag{3.100}$$

where $M(h)$ is its local martingale part, $B(h)$ is its finite variation part. Denote the continuous local martingale part of X by X^c.

Definition 3.12. (B, C, ν) is called as predictable characteristics of X if

(1) B is a predictable finite variation process, namely the process $B = B(h)$ appearing in (3.100);

(2) $C = < X^c, X^c >$;

(3) ν is the compensator of the random measure μ^X associated to the jumps of X.

Usually,

$$\widetilde{C} = < M(h), M(h) >$$

is called as the modified second characteristic of X. It can be obtained by

$$\begin{aligned}\widetilde{C} &= C + (h^2) * \nu - \sum_{s \le \cdot} \left(\int_{-\infty}^{+\infty} h(x)\nu(\{s\} \times dx) \right)^2 \\ &= C + (h^2) * \nu - \sum_{s \le \cdot} (\Delta B_s)^2.\end{aligned}$$

The following proposition is critical in the study of semimartingales.

Proposition 3.1. *There is equivalence between*

(i) X is a semimartingale with predictable characteristics (B, C, ν).

(ii) The following processes are local martingale

(a) $M(h) = X(h) - B - X_0$;

(b) $M(h)^2 - \widetilde{C}$;

(c) $g * \mu^X - g * \nu$ for any bounded Borel function g.

The following two inequalities are widely used in the study of limit theory of martingales.

Proposition 3.2. *(Lendlart inequality) Let X be a càdlág adapted process, A be the compensator of X, for any stopping times T and every $\varepsilon, \eta > 0$,*

$$\mathbb{P}(\sup_{s \le T} |X_s| \ge \varepsilon) \le \frac{\eta}{\varepsilon} + \mathbb{P}(A_T \ge \eta).$$

Proposition 3.3. *There exist two constants K_1 and K_2 such that every locally square-integrable martingale M with $M_0 = 0$ satisfies*

$$\mathbb{E}(\sup_{s \le t} M_s^4) \le K_1 a^2 \mathbb{E}(< M, M >_t^2)^{1/2} + K_2 \mathbb{E}(< M, M >_t^2),$$

where $a = \sup_{t,\omega} |\Delta M_t(\omega)|$.

Chapter 4

Convergence of Empirical Processes

If $X_1, \cdots, X_n$ are i.i.d. real valued random variables with common distribution function F, the empirical distribution function is

$$F_n(x) = \frac{1}{n}\sum_{i=1}^{n} 1_{\{X_i \le x\}}$$

and the corresponding *empirical process* is

$$Y_n(x) = \sqrt{n}(F_n(x) - F(x)).$$

The theory of empirical process is not only important in probability theory, but also provides a basis for studying statistics. In this chapter, we concern the weak convergence of empirical processes. In Section 1, we present the classical weak convergence result. As an application of theorems 3.6 and 4.1, weak convergence of marked empirical processes is given in Section 2. In Sections 3 and 4, the function index and set index empirical processes are studied.

4.1 Classical weak convergence of empirical processes

By the classical central limit theorem, for every fixed x, $Y_n(x)$ converge in distribution to a normal random variable with mean 0 and variance $F(x)(1 - F(x))$.

In 1952, Donsker proved a general extension for central limit theorem for $\{Y_n(x)\}$. He showed that the weak convergence of $\{Y_n\}$ to the Brownian bridge, B, holds for uniform $[0, 1]$ sample with respect to local uniform topology. Brownian bridge B is a Gaussian process satisfying

$$\mathbb{E}B_t = 0, \quad \mathbb{E}B_s B_t = s(1-t) \quad (0 \le s < t \le 1).$$

In fact, there exists a Wiener process W such that

$$B_t = W_t - tW_1.$$

However Donsker's formulation was not quite correct because of the problem of measurability of the functionals of discontinuous processes.

In 1956, Skorokhod introduced the Skorokhod topology, and proved that weak convergence under Skorokhod topology to a continuous process is equivalent to convergence under local uniform topology.

In this section, we introduce the classical weak convergence of empirical processes. For simplicity, we assume that $X_1, \cdots, X_n$ are i.i.d. real valued random variables with uniform distribution between 0 and 1. For $x \in [0,1]$,

$$Z_n(x) = \sqrt{n}(F_n(x) - x)$$

is random element on $\mathbb{D}[0,1]$. In order to discuss the weak convergence of $Z_n(x)$ under the local uniform topology, it is necessary to modify the random elements into $\mathbb{C}[0,1]$. Let $X_1^*, \cdots, X_n^*$ be the order statistics of sample $X_1, \cdots, X_n$, $X_0^* = 0$, $X_{n+1}^* = 1$, $G_n(X_i^*) = i/(n+1)$, and $G_n(x)$, $0 \le x \le 1$, the linear interpolation of $G_n(X_i^*)$. Then

$$G_n(x) = \frac{1}{n+1}\Big(\sum_{i=1}^n 1_{\{X_i \le x\}} + \frac{t - \max(X_i; X_i \le x)}{\min(X_i; X_i > x) - \max(X_i; X_i \le x)}\Big).$$

Define

$$W_n(x) = \sqrt{n}(G_n(x) - x).$$

We can easily obtain

$$|G_n(X_i^*) - F_n(X_i^*)| = \Big|\frac{i}{n+1} - \frac{i}{n}\Big| \le \frac{1}{n},$$

thus

$$\sup_{0 \le x \le 1} |G_n(x) - F_n(x)| \le \frac{1}{n}$$

and

$$\sup_{0 \le x \le 1} |W_n(x) - Z_n(x)| \le \frac{1}{\sqrt{n}}. \tag{4.1}$$

The classical weak convergence of empirical processes is presented in the following.

Theorem 4.1. *If $\{X_n\}_{n \ge 1}$ is a sequence of i.i.d. real valued random variables with uniform distribution between 0 and 1, we have*

$$Z_n \Rightarrow B \tag{4.2}$$

where B is a Brownian bridge.

Before proving this theorem, we first introduce a lemma.

Lemma 4.1. *Let $\{X_n\}_{n \ge 1}$ be a sequence of real valued random variables, $S_0 = 0$, $S_k = \sum_{i=1}^k X_i$, $M_n = \max_{0 \le k \le n} |S_k|$, $M_n' = \max_{0 \le k \le n} \min\{|S_k|, |S_n - S_k|\}$. If there are nonnegative real numbers $u_1, \cdots, u_n$, $\gamma \ge 0$ and $\alpha > 1/2$ such that*

$$\mathbb{P}(|S_j - S_i| \ge \lambda, |S_k - S_j| \ge \lambda) \le \frac{1}{\lambda^{2\gamma}}\Big(\sum_{i \le l \le k} u_l\Big)^{2\alpha} \tag{4.3}$$

for any $\lambda > 0$ *and* $0 \le i \le j \le k \le n$, *then*

$$\mathbb{P}(M'_n \ge \lambda) \le \frac{C_{\gamma,\alpha}}{\lambda^{2\gamma}}(\sum_{1\le l\le n} u_l)^{2\alpha}, \tag{4.4}$$

where $C_{\gamma,\alpha}$ *is a constant dependent on* γ, α.

Proof. Denote $\delta = (2\gamma+1)^{-1}$. For $K > 1$ large enough with

$$2(\frac{1}{K} + \frac{1}{2^{2\alpha}}) \le 1. \tag{4.5}$$

For such K, let $C_{\gamma,\alpha} = K$. We prove the lemma by mathematical induction method. When $n = 1$, (4.4) obviously holds.

For $n = 2$, $M'_2 = \min\{|S_1|, |S_2 - S_1|\}$,

$$\begin{aligned}\mathbb{P}(M'_2 \ge \lambda) &= \mathbb{P}(|S_1| \ge \lambda, |S_2 - S_1| \ge \lambda)\\ &\le \frac{1}{\lambda^{2\gamma}}(u_1+u_2)^{2\alpha} \le \frac{C_{\gamma,\alpha}}{\lambda^{2\gamma}}(u_1+u_2)^{2\alpha}\end{aligned}$$

by (4.3). We assume the lemma holds for integers which are less than n. Denote $u = \sum_{i=1}^n u_i$, we can assume $u > 0$. Then there exists integer h, such that $1 \le h \le n$, and

$$\frac{u_1 + \cdots + u_{h-1}}{u} \le \frac{1}{2} \le \frac{u_1 + \cdots + u_h}{u}.$$

Set

$$\begin{aligned}U_1 &= \max_{0\le i\le h-1} \min\{|S_i|, |S_{h-1} - S_i|\},\\ U_2 &= \max_{h\le j\le n} \min\{|S_j - S_h|, |(S_n - S_h) - (S_j - S_h)|\},\\ D_1 &= \min\{|S_{h-1}|, |S_n - S_{h-1}|\},\\ D_2 &= \min\{|S_h|, |S_n - S_h|\}.\end{aligned}$$

If $|S_i| \le U_1$,

$$\min\{|S_i|, |S_n - S_i|\} \le |S_i| \le U_1 \le U_1 + D_1;$$

If $|S_{h-1} - S_i| \le U_1$ and $|S_{h-1}| = D_1$,

$$\min\{|S_i|, |S_n - S_i|\} \le |S_i| \le |S_{h-1} - S_i| + |S_{h-1}| \le U_1 + D_1;$$

If $|S_i| \le U_1$ and $|S_n - S_{h-1}| = D_1$,

$$\min\{|S_i|, |S_n - S_i|\} \le |S_n - S_i| \le |S_{h-1} - S_i| + |S_n - S_{h-1}| \le U_1 + D_1.$$

Thus

$$\min\{|S_i|, |S_n - S_i|\} \le U_1 + D_1 \tag{4.6}$$

when $0 \le i \le h-1$. Similarly, we have

$$\min\{|S_i|, |S_n - S_i|\} \le U_2 + D_2 \tag{4.7}$$

when $h \le i \le n$. Hence

$$M_n' \le \min\{U_1 + D_1, U_2 + D_2\}. \tag{4.8}$$

From (4.8), for any given $\lambda > 0$

$$\mathbb{P}(M_n' \ge \lambda) \le \mathbb{P}(U_1 + D_1 \ge \lambda) + \mathbb{P}(U_2 + D_2 \ge \lambda).$$

Furthermore, let $\lambda = \lambda_1 + \lambda_2$,

$$\mathbb{P}(M_n' \ge \lambda) \le \mathbb{P}(U_1 \ge \lambda_1) + \mathbb{P}(U_2 \ge \lambda_1) + \mathbb{P}(D_1 \ge \lambda_2) + \mathbb{P}(D_2 \ge \lambda_2).$$

By the definition, $U_1 = M_{h-1}'$. Then by the induction assumption,

$$\mathbb{P}(U_1 \ge \lambda_1) \le \frac{C_{\gamma,\alpha}}{\lambda_1^{2\gamma}}(u_1 + \cdots + u_{h-1})^{2\alpha} \le \frac{u^{2\alpha}}{\lambda_1^{2\gamma}} \cdot \frac{C_{\gamma,\alpha}}{2^{2\alpha}}. \tag{4.9}$$

Similarly,

$$\mathbb{P}(U_2 \ge \lambda_1) \le \frac{C_{\gamma,\alpha}}{\lambda_1^{2\gamma}}(u_{h+1} + \cdots + u_n)^{2\alpha} \le \frac{u^{2\alpha}}{\lambda_1^{2\gamma}} \cdot \frac{C_{\gamma,\alpha}}{2^{2\alpha}}. \tag{4.10}$$

Furthermore,

$$\begin{aligned}\mathbb{P}(D_1 \ge \lambda_2) &= \mathbb{P}(|S_{h-1}| \ge \lambda_2, |S_n - S_{h-1}| \ge \lambda_2) \\ &\le \frac{1}{\lambda_2^{2\gamma}}(u_1 + \cdots + u_n)^{2\alpha} = \frac{u^{2\alpha}}{\lambda_2^{2\gamma}}.\end{aligned}$$

Similarly,

$$\mathbb{P}(D_2 \ge \lambda_2) \le \frac{u^{2\alpha}}{\lambda_2^{2\gamma}}.$$

Thus

$$\begin{aligned}\mathbb{P}(M_n' \ge \lambda) &\le 2u^{2\alpha}\Big(\frac{1}{\lambda_2^{2\gamma}} + \frac{C_{\gamma,\alpha}}{2^{2\alpha}\lambda_1^{2\gamma}}\Big) \\ &\le \frac{u^{2\alpha}}{\lambda^{2\gamma}} \cdot 2[(C_{\gamma,\alpha}2^{-2\alpha})^{(2\gamma+1)^{-1}} + 1]^{(2\gamma+1)} \le \frac{C_{\gamma,\alpha}}{\lambda^{2\gamma}}u^{2\alpha}.\end{aligned}$$

□

The proof of Theorem 4.1. Firstly, we prove the convergence of finite dimensional distribution of $\{Z_n\}$. Denote $U_n(t) = nF_n(t)$. For $0 = t_0 < t_1 < \cdots < t_k = 1$, let $p_i = t_i - t_{i-1}$, $U_n(t_i) - U_n(t_{i-1})$, $1 \le i \le k$, obeys multinomial distribution with parameters n and $p_1, \cdots, p_k$. By the central limit theorem of multinomial experiment,

$$(Z_n(t_i) - Z_n(t_{i-1}); 1 \le i \le k) \xrightarrow{d} (B(t_i) - B(t_{i-1}); 1 \le i \le k).$$

Secondly, we prove that $\{Z_n\}$ is tight. It is difficult to discuss the tightness of $\{Z_n\}$ on $\mathbb{D}[0,1]$, but it is easy to study the tightness of $\{W_n\}$ on $\mathbb{C}[0,1]$. In fact, we can focus on $\{W_n\}$ on $\mathbb{C}[0,1]$ by (4.1). We need to prove that for every $\varepsilon > 0$, $\eta > 0$, there exist δ, $0 < \delta < 1$ and n_0, such that

$$\mathbb{P}(\sup_{|s-t|\le\delta} |W_n(t) - W_n(s)| \ge \varepsilon) \le \delta\eta \tag{4.11}$$

as $n \geq n_0$, which can be implied by

$$\mathbb{P}(\sup_{0 \leq s \leq \delta} |Z_n(s)| \geq \varepsilon) \leq \delta\eta \tag{4.12}$$

as $n \geq n_0$. For fixed δ, let $\gamma = 2$, $\alpha = 1$, $u_i = \frac{\sqrt{6}\delta}{m}$. Apply $Z_n(i\delta/m) - Z_n((i-1)\delta/m)$ to Lemma 4.1. If we can prove

$$\mathbb{E}(|Z_n(s+p_1) - Z_n(s)|^2 |Z_n(s+p_1+p_2) - Z_n(s+p_1)|^2) \leq 6p_1p_2, \tag{4.13}$$

then

$$\mathbb{P}(\max_{1 \leq i \leq m} \min\{|Z_n(i\delta/m)|, |Z_n(\delta) - Z_n(i\delta/m)|\} \geq \varepsilon) \leq \frac{6C_{2,1}\delta^2}{\varepsilon^4}$$

by Lemma 4.1. Thus, we have

$$\mathbb{P}(\max_{1 \leq i \leq m} \{|Z_n(i\delta/m)|\} \geq \varepsilon) \leq \frac{96C\delta^2}{\varepsilon^4} + \mathbb{P}(|Z_n(\delta)| \geq \frac{\varepsilon}{2}).$$

Noticing

$$\lim_{m \to \infty} \mathbb{P}(\max_{1 \leq i \leq m} \{|Z_n(i\delta/m)|\} \geq \varepsilon) = \mathbb{P}(\sup_{0 \leq s \leq \delta} |Z_n(s)| \geq \varepsilon),$$

we have

$$\mathbb{P}(\sup_{0 \leq s \leq \delta} |Z_n(s)| \geq \varepsilon) \leq \frac{96C\delta^2}{\varepsilon^4} + \mathbb{P}(|Z_n(\delta)| \geq \frac{\varepsilon}{2}).$$

Furthermore,

$$\begin{aligned} \mathbb{P}(|Z_n(\delta)| \geq \frac{\varepsilon}{2}) &\to \mathbb{P}(N \geq \varepsilon/(2\sqrt{\delta(1-\delta)})) \\ &\leq \frac{16\delta^2}{\varepsilon^4}\mathbb{E}N \end{aligned}$$

since the central limit theorem for $Z_n(\delta)$, where N is standard normal distribution random variable. Hence

$$\mathbb{P}(\sup_{0 \leq s \leq \delta} |Z_n(s)| \geq \varepsilon) \leq (96C + 48)\frac{\delta^2}{\varepsilon^4}.$$

We can choose δ such that (4.12) holds.

Now we prove (4.13). Let $p = t - s$, we have

$$\begin{aligned} Z_n(t) - Z_n(s) &= \frac{1}{\sqrt{n}}(U_n(t) - U_n(s) - np) \\ &= \frac{1}{\sqrt{n}}(k - np) \end{aligned}$$

where k is the number of $X_1, \cdots, X_n$, which drop into $(s, t]$.

Define

$$\alpha_i = \begin{cases} 1 - p_1, & \text{when } X_i \in (s, s+p_1], \\ -p_1, & \text{otherwise}; \end{cases}$$

$$\beta_i = \begin{cases} 1 - p_2, & \text{when } X_i \in (s + p_1, s + p_1 + p_2], \\ -p_2, & \text{otherwise}; \end{cases}$$

$\{(\alpha_i, \beta_i), i = 1, \cdots, n\}$ is i.i.d. sequence with mean 0, and

$$\mathbb{E}[(\sum_{i=1}^n \alpha_i)^2 (\sum_{i=1}^n \beta_i)^2] = n\mathbb{E}\alpha_1^2\beta_1^2 + n(n-1)\mathbb{E}\alpha_1^2\mathbb{E}\beta_1^2$$
$$+2n(n-1)\mathbb{E}\alpha_1\beta_1\mathbb{E}\alpha_2\beta_2.$$

Noticing

$$\mathbb{E}\alpha_1^2\beta_1^2 = p_1(1-p_1)^2 p_2^2 + p_1^2(1-p_2)^2 p_2 + p_1^2 p_2^2(1 - p_1 - p_2) \le 3p_1p_2,$$
$$\mathbb{E}\alpha_1^2\mathbb{E}\beta_1^2 = p_1(1-p_1)p_2(1-p_2) \le p_1p_2,$$
$$\mathbb{E}\alpha_1\beta_1\mathbb{E}\alpha_2\beta_2 = p_1^2p_2^2 \le p_1p_2,$$

we obtain

$$\mathbb{E}[(\sum_{i=1}^n \alpha_i)^2 (\sum_{i=1}^n \beta_i)^2] \le 6n^2 p_1 p_2,$$

which implies (4.13). We complete the proof of Theorem 4.1.

4.2 Weak convergence of marked empirical processes

Consider the autoregressive model

$$X_i = \beta X_{i-1} + \varepsilon_i \tag{4.14}$$

where X_0 is given, $\{\varepsilon_i = G(\eta_i, \eta_{i-1}, \cdots)\}$ is a causal process with mean zero. The estimation and inference of β is a very interesting problem. The skills from empirical process and goodness-of-fit tests are vibrant research topic in statistics. In this section, we focus on the weak convergence results, which can be used in the inference of β.

Let $g(x)$ be a continuous function on $\mathbb{R}$. We assume $\{X_n\}_{n\ge1}$ is a unit root process in this section, thus there exists a constant sequence $\{a_n\}$ such that

$$\frac{X_{[n\cdot]}}{a_n} \Rightarrow \xi(\cdot)$$

for some stochastic process ξ, where ξ is a Lévy process or a Gaussian process. In this section, we study the weak convergence of

$$\alpha_n(x) = \sum_{i=1}^n g(\frac{X_i}{a_n})(1_{\{\varepsilon_i \le x\}} - F(x)) \tag{4.15}$$

and

$$\hat{\alpha}_n(x) = \sum_{i=1}^n g(\frac{X_i}{a_n})(1_{\{\hat{\varepsilon}_i \le x\}} - F(x)), \tag{4.16}$$

where $F(x)$ is the common distribution function of ε_i, $\hat{\varepsilon}_i = X_i - \hat{\beta}X_{i-1}$, and $\hat{\beta}$ is an estimate of β. $\alpha_n(x)$ and $\hat{\alpha}_n(x)$ are the so-called *marked empirical processes*. The results in this section are from *Chan and Zhang (2012)*.

Let

$$S_{[nt]} = \sum_{i=1}^{[nt]} \varepsilon_i,$$

and we assume ε_1 satisfies

$$\lim_{x\to\infty} \frac{\mathbb{P}(|\varepsilon_1| \geq xy)}{\mathbb{P}(|\varepsilon_1| \geq x)} = y^{-\alpha} \tag{4.17}$$

for any $y > 0$, where $\alpha \in (0, 2)$, and the normalization constants $\{a_n\}$ are given by

$$a_n = \inf\{x : \mathbb{P}(|\varepsilon_1| \geq x) \leq \frac{1}{n}\}.$$

In fact, condition (4.17) is equivalence to condition (2.66) in Chapter 2.

To study the asymptotic behavior of $\alpha_n(x)$ and $\hat{\alpha}_n(x)$, we utilize the martingale approximation. Let

$$\begin{aligned}
&\mathcal{F}_i = \sigma(\eta_i, \eta_{i-1}, \cdots), \quad \mathcal{F}_i^* = \sigma(\eta_i, \eta_{i-1}, \cdots, \eta_1, \eta_0', \eta_{-1}, \cdots),\\
&F_i(x|\mathcal{F}_j) = \mathbb{P}(\varepsilon_i \leq x|\mathcal{F}_j), \quad f_i(x|\mathcal{F}_j) = F_i'(x|\mathcal{F}_j),\\
&\mathcal{P}_j(\cdot) = \mathbb{E}(\cdot|\mathcal{F}_j) - \mathbb{E}(\cdot|\mathcal{F}_{j-1}),
\end{aligned}$$

where η_0' is the independent copy of η_0. Let

$$B_n(t, x) = \sum_{i=1}^{[nt]} [1_{\{\varepsilon_i < x\}} - F(x)]$$

and $B(t, x)$ be a rescaled Brownian bridge for fix t and a Brownian motion with variance $\mu(x) = \mathbb{E}\{\sum_{i=0}^{\infty} F_1(x|\mathcal{F}_0) - F_i(x|\mathcal{F}_0^*)\}^2$ for fix x.

Theorem 4.2. *Let $g(\cdot)$ be a Lipschitz continuous function on $\mathbb{R}$, i.e., $|g(x) - g(y)| \leq C|x - y|$ for all $x, y \in (-\infty, +\infty)$. Assume that there exist a_n such that*

$$\frac{S_{[n\cdot]}}{a_n} \Rightarrow S(\cdot) \tag{4.18}$$

on $\mathbb{D}[0, 1]$ with the semimartingale topology, where S is a Lévy process or a Gaussian process, and

$$\sum_{j=1}^{\infty} || \sum_{i=j}^{\infty} F_i(x|\mathcal{F}_0) - F_i(x|\mathcal{F}_0^*)||^2 < \infty. \tag{4.19}$$

Then for any $x \in \mathbb{R}$,

$$\frac{1}{\sqrt{n}} \alpha_n(x) \xrightarrow{d} \int_0^1 g(S(t-))dB(t, x). \tag{4.20}$$

If (4.19) is replaced by

$$\sum_{j=1}^{\infty} \sup_x |\sum_{i=j}^{\infty} F_i^{(l)}(x|\mathcal{F}_0) - F_i^{(l)}(x|\mathcal{F}_0^*)|^2 < \infty, \tag{4.21}$$

where $l = 0, 1$, $F_i^{(0)}(x|\mathcal{F}_0) = F_i(x|\mathcal{F}_0)$, *then for any constant* $A > 0$,

$$\frac{1}{\sqrt{n}} \alpha_n(\cdot) \Rightarrow \int_0^1 g(S(t-)) dB(t, \cdot) \tag{4.22}$$

in $\mathbb{D}[-A, A]$ *under the* J_1 *topology.*

Theorem 4.3. *If* $a_n(\hat{\beta} - 1) = o_{\mathbb{P}}(1)$, *then*

$$\frac{1}{\sqrt{n}} \hat{\alpha}_n(x) = \frac{1}{\sqrt{n}} \alpha_n(x) + \frac{1}{\sqrt{n}} \sum_{i=1}^{n} g(X_{i-1}/a_n)[F(x + (\hat{\beta} - \beta)X_{i-1}) - F(x)] + o_{\mathbb{P}}(1). \tag{4.23}$$

Let $\hat{\beta}$ be the τ-quantile estimate of β, that is,

$$\hat{\beta} = \arg\min_{\beta} \sum_{i=1}^{n} \rho_\tau (X_i - \beta X_{i-1} - F^{-1}(\tau)),$$

where $\rho_\tau(y) = y(\tau - 1_{\{y \le 0\}})$. When $\beta = 1$, using the argument of Theorem 4 in Knight (1991), we have

$$a_n \sqrt{n} (\hat{\beta} - \beta) = \frac{\frac{1}{\sqrt{n}} \sum_{t=1}^{n} (X_{t-1}/a_n)(\tau - 1_{\{\varepsilon_t \le F^{-1}(\tau)\}})}{\frac{1}{n} \sum_{t=1}^{n} f_t(F^{-1}(\tau)|\mathcal{F}_{t-1})(X_{t-1}^2/a_n^2)} + o_{\mathbb{P}}(1).$$

By virtue of Theorem 2.2 and this expression, the following corollary concerning the quantile estimate is immediate. Let f denote the density function of ε_1.

Corollary 4.1. *Under the conditions in Theorem 4.2, if* $\mathbb{E}|f_1(F^{-1}(\tau)|\mathcal{F}_0|)^p < \infty$ *for some* $p > 1$ *and* $f(F^{-1}(\tau)) > 0$, *then*

$$a_n \sqrt{n} (\hat{\beta} - \beta) \xrightarrow{d} -\frac{1}{f(F^{-1}(\tau))} \frac{\int_0^1 S(t-) dB(t, F^{-1}(\tau))}{\int_0^1 S^2(t) dt}.$$

Theorem 4.4. *Under the conditions in Theorem 4.2 we have*

(1) if $\hat{\beta}$ *is the* τ*-quantile estimate of* β *and* $f(F^{-1}) > 0$, *then*

$$\sup_{x \in [-A, A]} \frac{1}{\sqrt{n}} \hat{\alpha}_n(x)$$

$$\xrightarrow{d} \sup_{x \in [-A, A]} [(-\frac{f(x)}{f(F^{-1}(\tau))})(\frac{\int_0^1 S(t-) dB(t, F^{-1}(\tau))}{\int_0^1 S^2(t) dt}) \int_0^1 g(S(t) S(t) dt$$

$$+ \int_0^1 g(S(t-)) dB(t, x)]$$

(2) if $\hat{\beta}$ is the least square estimation (LSE) of β and

$$(S_{[nt]}, \sum_{i=1}^{n} \varepsilon_i^2/a_n^2) \xrightarrow{d} (S(t), S^2),$$

then

(i) if $a_n = n^{\vartheta} l(n)$ for some $1/2 < \vartheta < 1$,

$$\sup_{x\in[-A,A]} \frac{1}{\sqrt{n}} \hat{\alpha}_n(x) \xrightarrow{d} \sup_{x\in[-A,A]} f(x) \int_0^1 S(t-)dS(t) \int_0^1 g(S(t))S(t)dt / \int_0^1 S^2(t);$$

(ii) if $a_n = \sqrt{n}$,

$$\sup_{x\in[-A,A]} \frac{1}{\sqrt{n}} \hat{\alpha}_n(x) \xrightarrow{d} \sup_{x\in[-A,A]} [f(x) \int_0^1 S(t-)dS(t) \int_0^1 g(S(t))S(t)dt / \int_0^1 S^2(t) + \int_0^1 g(S(t-))dB(t,x)].$$

Before proving these results, it is necessary to introduce some lemmas.

Lemma 4.2. *(Wu (2003)) Let H be a function with continuous first-order derivatives and $a > 0$. Then*

$$\sup_{t\leq s\leq t+a} H^2(s) \leq \frac{2}{a} \int_t^{t+a} H^2(u)du + 2a \int_t^{t+a} H'^2(u)du$$

and

$$\sup_{s\in\mathbb{R}} H^2(s) \leq 2 \int_{\mathbb{R}} H^2(u)du + 2 \int_{\mathbb{R}} H'^2(u)du,$$

where H' is the derivative of H.

Lemma 4.3. *Under the conditions in Theorem 4.2, for any x, there exists a martingale difference sequence $\zeta_i(x)$ with respect to $\mathcal{F}_i$ such that for any $\delta > 0$,*

$$\lim_{n\to\infty} \mathbb{P}\{|\frac{1}{\sqrt{n}} \sum_{i=1}^{n} g(S_{i-1}/a_n)(1_{\{\varepsilon_i\leq x\}} - F(x)) - \frac{1}{\sqrt{n}} \sum_{i=1}^{n} g(S_{i-1}/a_n)\zeta_i(x)| > \delta\} = 0.$$

Proof. When (4.19) holds, by Volny (1993), there exists a random variables sequence

$$\xi_i(x) = \sum_{j=-\infty}^{-1} \sum_{l=0}^{\infty} \mathcal{P}_{i+j} 1_{\{\varepsilon_{i+l}\leq x\}}$$

and a martingale difference sequence

$$\zeta_i(x) = \sum_{j=-i}^{\infty} \mathcal{P}_i 1_{\{\varepsilon_j\leq x\}}$$

such that

$$1_{\{\varepsilon_i\leq x\}} - F(x) = \zeta_i(x) + \xi_i(x) - \xi_{i+1}(x).$$

This gives that

$$\frac{1}{\sqrt{n}}\sum_{i=1}^{n} g(S_{i-1}/a_n)(1_{\{\varepsilon_i\le x\}}-F(x))-\frac{1}{\sqrt{n}}\sum_{i=1}^{n} g(S_{i-1}/a_n)\zeta_i(x)$$

$$=\frac{1}{\sqrt{n}}\sum_{i=1}^{n-1}\xi_{i+1}(x)[g(S_i/a_n)-g(S_{i-1}/a_n)]-\frac{1}{\sqrt{n}}g(S_{n-1}/a_n)\xi_{n+1}(x)=:I_1+I_2.$$

Since for any $\delta>0$,

$$\mathbb{P}\{\sup_{2\le i\le n+1}|\xi_i(x)|>\delta\sqrt{n}\}\le\sum_{i=2}^{n+1}(\sqrt{n}\delta)^{-2}\mathbb{E}[\xi_1^2(x)I(|\xi_1(x)|>\delta\sqrt{n})]\to 0,$$

it follows that $|\xi_{n+1}(x)|/\sqrt{n}=o_{\mathbb{P}}(1)$.

On the other hand, by (4.18),

$$g(S_{n-1}/a_n)=O_{\mathbb{P}}(1)$$

holds. Thus, $I_2=o_{\mathbb{P}}(1)$. It suffices to show that $I_1=o_{\mathbb{P}}(1)$.

When $\{\varepsilon_i\}$ has infinite variance, the result $I_1=o_{\mathbb{P}}(1)$ follows along exactly the lines of argument of Lemma 2 of Knight (1991). We therefore only give the proof for the finite variance case in detail.

When $\{\varepsilon_i\}$ has finite variance, by Theorem 1 of Wu (2007), $\mathbb{E}(\sum_{i=1}^{n}\varepsilon_i)^2=O(\sqrt{n})$. Thus, $a_n=O(\sqrt{n})$. By the assumptions, we have

$$\begin{aligned}\mathbb{E}|I_1|&\le\frac{C}{\sqrt{n}}\sum_{i=1}^{n-1}\mathbb{E}|\xi_{i+1}(x)\varepsilon_i/a_n|\\&\le\frac{C'}{n}\sum_{i=1}^{n-1}[\mathbb{E}\xi_{i+1}^2(x)]^{1/2}[\mathbb{E}\varepsilon_i^2]^{1/2}\le\frac{C''}{n}\sum_{i=2}^{n}[\mathbb{E}\xi_i^2(x)]^{1/2}\\&\le\frac{C'''}{n}\sum_{i=2}^{n}\{\sum_{j=-\infty}^{-1}\mathbb{E}[\sum_{l=0}^{\infty}[F_{i+l}(x|\mathcal{F}_j)-F_{i+l}(x|\mathcal{F}_j^*)]]^2\}^{1/2}=o(1).\end{aligned}$$

Then $I_1=o_{\mathbb{P}}(1)$ and therefore the proof is complete. □

Lemma 4.4. *Under the conditions in Theorem 4.2, for any constant $A>0$,*

$$\sup_{x\in[-A,A]}|\frac{1}{\sqrt{n}}\sum_{i=1}^{n} g(S_{i-1}/a_n)(1_{\{\varepsilon_i<x\}}-F(x))-\frac{1}{\sqrt{n}}\sum_{i=1}^{n} g(S_{i-1}/a_n)\zeta_i(x)|$$

converges to zero in probability, where $\zeta_i(x)$ is defined as in Lemma 4.3.

Proof. It suffices to show

$$\frac{1}{n}\sum_{i=1}^{n}\mathbb{E}\sup_{x\in[-A,A]}\xi_i^2(x)=O(1).$$

Since $\xi_i(x) = \sum_{j=-\infty}^{-1} \sum_{l=0}^{\infty} \mathcal{P}_j 1_{\{\varepsilon_{i+l} \le x\}}$ is differentiable, thus

$$\begin{aligned}
\mathbb{E} \sup_{x \in [-A,A]} \xi_i^2(x) &\le \frac{2}{A} \int_{-A}^{A} \mathbb{E}\xi_i^2(u)du + 2A \int_{-A}^{A} \mathbb{E}\xi_i'^2(u)du \\
&\le \frac{2}{A} \int_{-A}^{A} \sum_{j=-\infty}^{-1} \mathbb{E}[\sum_{l=0}^{\infty}[F_{i+l}(u|\mathcal{F}_j) - F_{i+l}(u|\mathcal{F}_j^*)]]^2 du \\
&\quad +2A \int_{-A}^{A} \sum_{j=-\infty}^{-1} \mathbb{E}[\sum_{l=0}^{\infty}[f_{i+l}(u|\mathcal{F}_j) - f_{i+l}(u|\mathcal{F}_j^*)]]^2 du \\
&\le 2 \sum_{j=-\infty}^{-1} \sup_u \mathbb{E}[\sum_{l=0}^{\infty}[F_{i+l}(u|\mathcal{F}_j) - F_{i+l}(u|\mathcal{F}_j^*)]]^2 \\
&\quad +4A^2 \sum_{j=-\infty}^{-1} \sup_u \mathbb{E}[\sum_{l=0}^{\infty}[f_{i+l}(u|\mathcal{F}_j) - f_{i+l}(u|\mathcal{F}_j^*)]]^2 \\
&\le C
\end{aligned}$$

for every i. □

Lemma 4.5. *Let*

$$\tilde{B}_n(t,x) = \sum_{i=1}^{[nt]} \zeta_i(x),$$

$\zeta_i(x)$ is the martingale difference defined in Lemma 4.3. Then under condition (4.21),

$$\frac{1}{\sqrt{n}} \tilde{B}_n(\cdot,\cdot) \Rightarrow B(\cdot,\cdot)$$

in $\mathbb{D}([0,1] \times [-A,A])$.

Proof. (4.21) implies

$$\sum_{j=1}^{\infty} || \sum_{i=j}^{\infty} F_i(x|\mathcal{F}_0) - F_i(x|\mathcal{F}_0^*)||^2 < \infty,$$

it follows that

$$\mathbb{E}\zeta_i(x) = \mathbb{E}\{\sum_{i=0}^{\infty} F_i(x|\mathcal{F}_0) - F_i(x|\mathcal{F}_0^*)\}^2 = \mu(x) < \infty.$$

Since $\{\zeta_i(x)\}$ is a martingale difference sequence, by Theorem 18.2 of Billingsley (1999), we have

$$\frac{1}{\sqrt{n}} \tilde{B}_n(\cdot,\cdot) \Rightarrow B(\cdot,\cdot).$$

Thus, the finite dimensional convergence of $\tilde{B}_n(t,x)$ follows. Let $\zeta_i(x,y) = \zeta_i(y) - \zeta_i(x)$. By Theorem 6 of Bickel and Wicchura (1971), to show the tightness of

$\{\tilde{B}_n(t,x)\}$ on $\mathbb{D}[0,1]\times\mathbb{D}[-A,A]$, it suffices to show that for any $0\le t_1<t<t_2\le 1$ and $-A\le x_1<x<x_2\le A$,

$$n^{-2}\mathbb{E}\{[\sum_{i=[nt_1]+1}^{[nt]}\zeta_i(x_1,x_2)]^2[\sum_{i=[nt]+1}^{[nt_2]}\zeta_i(x_1,x_2)]^2\}\le(t-t_1)(t_2-t)(x_2-x_1)^2$$

and

$$\mathbb{E}\{|\sum_{i=[nt_1]+1}^{[nt_2]}\zeta_i(x_1,x_2)|^2|\sum_{i=[nt_1]+1}^{[nt_2]}\zeta_i(x_1,x_2)|^2\}\le C(x-x_1)(x_2-x)(t_2-t_1)^2.$$

These follow easily by condition (4.21) and noting that $\tilde{B}_n(t,x)$ is a martingale. Details are omitted. □

Lemma 4.6. *Under the conditions of Theorem 4.2, there exists a dense set $Q\subset[0,1]$,$0,1\in Q$ such that for any finite subset $\{0\le t_1<t_2<\cdots<t_m\le 1\}\subset Q$ and for any x,*

$$(S_{[nt_i]},\tilde{B}_n(t_i,x),1\le i\le m)\xrightarrow{d}(S(t_i),B(t_i,x),1\le i\le m).\tag{4.24}$$

Proof. There exists a dense set $Q^{'}\subset[0,1],1\in Q^{'}$ such that for any finite subset $\{t_1<t_2<\cdots<t_m\le 1\}\subset Q^{'}$,

$$(S_{[nt_1]},S_{[nt_2]},S_{[nt_3]})\xrightarrow{d}(S(t_1),S(t_2),S(t_3))$$

by (4.18).

When ε_i has infinite variance, weak convergence in $\mathbb{D}([0,1]\times[-A,A])$ in Lemma 4.5 can also be replaced by that in $\mathbb{C}([0,1]\times[-A,A])$, since $W(t,x)$ is a continuous process on $[0,1]\times[-A,A]$. Thus $(S_{[nt]},\tilde{B}_n(t,x))$ is uniformly S-tight on $\mathbb{D}([0,1]\times[-A,A])$. This implies that for sequence $(S_{[nt]},\tilde{B}_n(t,x)),t\in Q$, there exists a subsequence $(S_{[n_kt]},\tilde{B}_{n_k}(t,x))$ such that

$$(S_{[n_kt]},\tilde{B}_{n_k}(t,x))\xrightarrow{d}(S(t),B(t,x)).$$

Following the argument of Theorem 3 in Resnick and Greenwood (1979), we have that $S(t)$ and $B(t,x)$ are independent and any convergent subsequence has the same limit. Thus, (4.24) holds.

When ε_i has finite variance, from Corollary 3 of Dedecker and Merlevède (2003), $S_{[n\cdot]}\Rightarrow S(\cdot)$. Thus, by Lemma 4.5, if we can show that for any finite subset $\{t_i,1\le t\le m\}\subset[0,1]$,

$$(S_{[nt_i]},\tilde{B}_n(t_i,x),1\le i\le m)\xrightarrow{d}(S(t_i),B(t_i,x),1\le i\le m)\tag{4.25}$$

then

$$(S_{[n\cdot]},\tilde{B}_n(\cdot,x))\Rightarrow(S(\cdot),B(\cdot,x))$$

on $\mathbb{D}[0,1]$ and (4.24) follows. By Theorem 1 of Wu (2007), there exists martingale $\mathbb{M}_i$ with respect to $\mathcal{F}_i$ such that

$$|(S_n(t), \tilde{B}_n(t,x)) - (\sum_{i=1}^{[nt]} \mathbb{M}_i, \sum_{i=1}^{[nt]} \zeta_i(x))| = o_{\mathbb{P}}(1).$$

On the other hand, from the martingale central limit theorem, it follows that

$$\sum_{i=1}^{[n\cdot]} (\mathbb{M}_i, \zeta_i(x)) \Rightarrow (S(\cdot), B(\cdot, x)).$$

Thus, (4.25) holds, we obtain the lemma. □

The proof of Theorem 4.2.

Lemma 4.6 implies that (5) of Jakubowski (1996) holds, i.e., there exists a dense set Q such that for any x

$$(g(S_{[nt]}), \tilde{B}_n(t,x)) \xrightarrow{d} (g(S(t)), B(t,x)), \quad t \in Q.$$

Further, since $g(\cdot)$ is a Lipschitz continuous function and $S_{[n\cdot]}$ is uniformly S-tight, it follows that $g(S_{[n\cdot]})$ is also uniformly S-tight. Moreover, for any $x \in \mathbb{R}$, $\tilde{B}(t,x)$ is a martingale satisfying UT condition and is J_1-tight with limiting law concentrated on $\mathbb{C}([0,1])$, by Remark 4 of Jakubowski (1996), we see that his condition (6) is satisfied.

Therefore, for $\{g(S_{[n\cdot]}), \tilde{B}_n(\cdot, x)\}$, all the conditions of Theorem 3 of Jakubowski (1996) are satisfied, thus

$$\frac{1}{\sqrt{n}} \sum_{i=1}^{n} g(S_{i-1}/a_n)\zeta_i(x) \xrightarrow{d} \int_0^1 g(S(t-))dB(t,x). \tag{4.26}$$

The left side of (4.20) can be approximated by $\frac{1}{\sqrt{n}} \sum_{i=1}^{n} g(S_{i-1}/a_n)\zeta_i(x)$ through martingale approximation. Thus, (4.20) follows from Lemma 4.3.

By Lemma 4.4, for (4.22) it suffices to show that

$$U_n(\cdot) := \frac{1}{\sqrt{n}} \sum_{i=1}^{n} g(S_{i-1}/a_n)\zeta_i(\cdot) \Rightarrow \int_0^1 g(S(t-))dB(t,\cdot) =: U(\cdot)$$

on $\mathbb{D}[0,1]$. The finite-dimension convergence follows from (4.26) and the Cramér-Wold device. Next, we show for any $\varepsilon > 0$, there exists a $\delta > 0$ such that

$$\mathbb{P}\{ \sup_{|x-y| \le \delta} |U_n(x) - U_n(y)| > \varepsilon\} \to 0 \tag{4.27}$$

as $n \to \infty$. This implies $\{U_n(\cdot)\}$ is tight, as a result, we complete the proof.

Obviously,

$$\sup_{0 \le t \le 1} |g(S_{[nt]})| \xrightarrow{d} \sup_{0 \le t \le 1} |g(S(t))|,$$

by (4.18). Let

$$g_\delta(S_i/a_n) = g(S_i/a_n) 1_{\{|g(S_i/a_n)| \le \delta^{-1/4}\}}$$

and

$$V_n(x) = \frac{1}{\sqrt{n}} \sum_{i=1}^{n} g_\delta(S_{i-1}/a_n)(\zeta_i(x) - \zeta_i(y)).$$

Then $V_n(x)$ is a martingale and by Lemma 4.2 and condition (4.21),

$$\begin{aligned}
&\mathbb{E}[\sup_{y\le x\le y+\delta} |V_n(x)|]^2 \\
&\le \frac{2}{\delta n}\int_y^{y+\delta} \mathbb{E}(\sum_{i=1}^{n}[g_\delta(\frac{S_{i-1}}{a_n})(\zeta_i(u)-\zeta_i(y))])^2 du + \frac{2\delta}{n}\int_y^{y+\delta} \mathbb{E}(\sum_{i=1}^{n}[g_\delta(\frac{S_{i-1}}{a_n})\zeta_i^{'}(u)])^2 du \\
&\le \frac{2}{\sqrt{\delta}n}\sum_{i=1}^{n}\int_y^{y+\delta} \mathbb{E}\{\zeta_i(u)-\zeta_i(y)\}^2 du + \frac{2\delta^2}{\sqrt{\delta}n}\sum_{i=1}^{n}\sup_{y\le x\le y+\delta} \mathbb{E}\{\zeta_i^{'}(x)\}^2 \\
&\le \frac{2}{\sqrt{\delta}n}\sum_{i=1}^{n}\int_y^{y+\delta}\int_y^{u} \mathbb{E}\{\zeta_i^{'}(v)\}^2 dv du + \frac{2\delta^2}{\sqrt{\delta}n}\sup_{x\in[-A,A]} \mathbb{E}\{\zeta_i^{'}(x)\}^2 \le C\delta^{3/2}.
\end{aligned}$$

Note that

$$\begin{aligned}
&\mathbb{P}\{\sup_{|x-y|\le\delta} |U_n(x) - U_n(y)| > 4\varepsilon\} \\
&\le (1+[1/\delta])\mathbb{P}\{\sup_{y\le x\le y+\delta} |V_n(x) > \varepsilon|\} + \mathbb{P}\{\max_{1\le i\le n} |g(S_i/a_n)| > \delta^{-1/4}\},
\end{aligned}$$

which implies (4.27). We complete the proof.

The proof of Theorem 4.3. Let $\{u_{ni}\}_{1\le i\le n}$ be a constant array with $\max_{1\le i\le n}|u_{ni}| = o(1)$. Along the lines of proof in Lemma 4.3, we have

$$\begin{aligned}
&\frac{1}{\sqrt{n}}\sum_{i=1}^{n} g(\frac{S_{i-1}}{a_n})[1_{\{\varepsilon_i\le x+u_{ni}\}} - 1_{\{\varepsilon_i\le x\}} \\
&= \frac{1}{\sqrt{n}}\sum_{i=1}^{n} g(\frac{S_{i-1}}{a_n})(\zeta_i(x+u_{ni}) - \zeta_i(x)) + \frac{1}{\sqrt{n}}\sum_{i=1}^{n} g(\frac{S_{i-1}}{a_n})(F(x+u_{ni}) - F(x)) + o_{\mathbb{P}}(1) \\
&= \frac{1}{\sqrt{n}}\sum_{i=1}^{n} g(\frac{S_{i-1}}{a_n})(F(x+u_{ni}) - F(x)) + o_{\mathbb{P}}(1).
\end{aligned}$$

If $a_n(\hat{\beta} - \beta) = o_{\mathbb{P}}(1)$,

$$\max_{1\le i\le n} (\hat{\beta} - \beta)X_i = o_{\mathbb{P}}(1),$$

since

$$\max_{1\le i\le n} |X_i/a_n| = O_{\mathbb{P}}(1).$$

Thus, we have

$$\frac{1}{\sqrt{n}}(\hat{\alpha}_n(x) - \alpha_n(x)) = \frac{1}{\sqrt{n}}\sum_{i=1}^{n} g(S_{i-1}/a_n)(F(x + (\hat{\beta} - \beta)X_i) - F(x)) + o_{\mathbb{P}}(1).$$

This completes the proof of Theorem 4.3.

The proof of Theorem 4.4.

Let $\{u_{ni}\}_{1\le i\le n}$ be a constant with $\max_i |u_{ni}| = o(1)$, then by Lemma 4.4, we have

$$\begin{aligned}
&\sup_{x\in[-A,A]} \frac{1}{\sqrt{n}}\sum_{i=1}^n g(X_{i-1}/a_n)1_{\{\varepsilon_i\le x+u_{ni}\}}\\
=& \sup_{x\in[-A,A]}[\frac{1}{\sqrt{n}}\sum_{i=1}^n g(X_{i-1}/a_n)1_{\{\varepsilon_i\le x\}} + \frac{1}{\sqrt{n}}\sum_{i=1}^n g(X_{i-1}/a_n)[\zeta_i(x+u_{ni}) - \zeta_i(x)]\\
&+\frac{1}{\sqrt{n}}\sum_{i=1}^n g(X_{i-1}/a_n)(F(x+u_{ni}) - F(x))] + o_{\mathbb{P}}(1)\\
=& \sup_{x\in[-A,A]}[\frac{1}{\sqrt{n}}\sum_{i=1}^n g(X_{i-1}/a_n)1_{\{\varepsilon_i\le x\}}\\
&+\frac{1}{\sqrt{n}}\sum_{i=1}^n g(X_{i-1}/a_n)(F_i(x+u_{ni}) - F_i(x))] + o_{\mathbb{P}}(1).
\end{aligned}$$

As a result, by $\max_{1\le i\le n}(\hat\beta - 1)X_i = o_{\mathbb{P}}(1)$ and the Taylors expansion, we have

$$\sup_{x\in[-A,A]}\frac{\hat\alpha_n(x)}{\sqrt{n}} = \sup_{x\in[-A,A]}[\frac{\alpha_n(x)}{\sqrt{n}} + \frac{1}{\sqrt{n}}f(x)(\hat\beta-1)\sum_{i=1}^n g(X_{i-1}/a_n)X_{i-1}] + o_{\mathbb{P}}(1)$$

in probability. Further, by Theorem 3 of Jakubowski (1996), it follows that

$$\frac{1}{\sqrt{n}}\sum_{i=1}^n[g(X_{i-1}/a_n)X_{i-1}/a_n] \xrightarrow{d} \int_0^1 g(S(t))(S(t))dt.$$

Corollary 4.1 yields (1).

As for (2), noting that $\hat\beta$ is the LSE of β, we have

$$\begin{aligned}
n(\hat\beta - \beta) &= \frac{1}{2}[X_n^2/a_n^2 - X_0^2/a_n^2 - \sum_{i=1}^n \varepsilon_i^2/a_n^2]/[\frac{1}{\sqrt{n}}\sum_{i=1}^n X_{i-1}^2/a_n^2]\\
&\xrightarrow{d} \frac{1}{2}(S^2(1) - S^2)/\int_0^1 S^2(t)dt =: \int_0^1 S(t-)dS(t)/\int_0^1 S^2(t)dt.
\end{aligned}$$

The proof of Theorem 4.4 is completed.

4.3 Weak convergence of function index empirical processes

In Section 4.1, we have already mentioned that Donsker's formulation was not quite correct since he discussed the weak convergence of empirical process on $\mathbb{D}[0,1]$ under the local uniform topology, however measurability of the functionals of discontinuous processes may not hold in this case. Skorokhod (1956) introduced the Skorokhod metric to deal with this problem, however, the Skorokhod metric is quite complex.

Dudley (1978), *Talagrand (1987)* and so on reformulated Donsker's result to avoid the problem of measurability and the need of the Skorokhod metric.

In this section, we will introduce this approach and discuss the weak convergence of function index empirical processes. Function index empirical processes can be seen as an extension of classical empirical processes, however, the tools of studying are quite different.

In the whole section, let $\{X_n\}_{n\geq 1}$ be a sequence of i.i.d. random variables defined on $(\Omega, \mathcal{F}, \mathbb{P})$, and P the common distribution of $\{X_n\}_{n\geq 1}$. The so called empirical measure of $\{X_n\}_{n\geq 1}$ is defined as:

$$P_n = \frac{1}{n}\sum_{i=1}^{n} \delta_{X_i},$$

where δ is the dirac measure at the observation.

Given a collection $\mathfrak{M}$ of $\mathbb{R} \to \mathbb{R}$ measurable functions we define a map from $\mathfrak{M}$ to $\mathbb{R}$ by

$$g \mapsto P_n(g), \qquad g \in \mathfrak{M},$$

here, we use the abbreviation $Q(g) = \int g dQ$ for measurable function g and measure Q. Then

$$P_n(g) = \int g dP_n = \frac{1}{n}\sum_{i=1}^{n} \int g d\delta_{X_i} = \frac{1}{n}\sum_{i=1}^{n} g(X_i).$$

4.3.1 *Preliminary*

The core of this section is to investigate weak convergence in the space of $l^\infty(\mathfrak{M})$, the set of all uniformly bounded real functions on $\mathfrak{M}$ with the norm

$$||z||_{\mathfrak{M}} = \sup_{g\in\mathfrak{M}} |z(g)|.$$

The metric space $(l^\infty(\mathfrak{M}), ||\cdot||_{\mathfrak{M}})$ is not a separable metric space. Thus, if f is a bounded continuous function on $l^\infty(\mathfrak{M})$ and ξ is a random element taking values in $l^\infty(\mathfrak{M})$, the measurability of $f(\xi)$ maybe fail. This is quite different from the cases in previous chapters. Hence it is necessary to modify the definition of weak convergence for nonmeasurable variables on $(\Omega, \mathcal{F}, \mathbb{P})$.

Let Z be an arbitrary map from $(\Omega, \mathcal{F}, \mathbb{P})$ to $\mathbb{R}$. The outer expectation of Z is defined as

$$\mathbb{E}^* Z = \inf\{\mathbb{E}U : U \geq Z, \ \ U : \Omega \to \mathbb{R} \ \text{ measurable and } \ \mathbb{E}U \text{ exists}\}.$$

Furthermore, the outer probability measure is defined as

$$\mathbb{P}^*(A) = \mathbb{E}^* 1_A$$

for every set $A \subset \Omega$.

Definition 4.1. Let $\{H_n\}$ be a sequence of arbitrary, possibly nonmeasurable maps on $(\Omega, \mathcal{F}, \mathbb{P})$, H be a Borel measurable map. If for every bounded continuous functional f,

$$\lim_{n\to\infty} \mathbb{E}^* f(H_n) = \mathbb{E} f(H),$$

we say $\{H_n\}$ converges weakly to H, and write $H_n \Rightarrow H$.

Function index empirical process is given by

$$\mathbb{G}_n(g) := P_n(g) - P(g) = \frac{1}{n}\sum_{i=1}^{n}(g(X_i) - P(g)), \tag{4.28}$$

where $g \in \mathfrak{M}$. The empirical process $\mathbb{G}_n$ can be viewed as a map from $\mathfrak{M}$ into $l^\infty(\mathfrak{M})$. Consequently, it makes sense to investigate conditions under which

$$\sqrt{n}\mathbb{G}_n \Rightarrow \mathbb{G} \tag{4.29}$$

in $l^\infty(\mathfrak{M})$, where $\mathbb{G}$ is a Gaussian random element in $l^\infty(\mathfrak{M})$.

Definition 4.2. A class $\mathfrak{M}$ for which (4.29) is true is called a *$P-$Donsker class.*

To prove (4.29) in the case of Definition 4.1, the results in Section 1.3 may not be enough, here we present a more general result.

Theorem 4.5. *If $\{H_n\}$ is a sequence of random elements taking values in $l^\infty(\mathfrak{M})$, and there exists a pseudometric $||\cdot||$ on $\mathfrak{M}$ such that $(\mathfrak{M}, ||\cdot||)$ is totally bounded and $\{H_n\}$ is asymptotic equicontinuity, i.e.,*

$$\lim_{\delta\to 0}\limsup_{n\to\infty} \mathbb{P}^*(\sup_{||g-h||\le\delta} |H_n(g) - H_n(h)| > \varepsilon) = 0 \tag{4.30}$$

for all $\varepsilon > 0$. Furthermore, all the finite-dimensional distributions of H_n converge in law to the corresponding finite-dimensional distributions of H, which has tight probability distribution on $l^\infty(\mathfrak{M})$. Then

$$H_n \Rightarrow H$$

in $l^\infty(\mathfrak{M})$.

Proof. Since $(\mathfrak{M}, ||\cdot||)$ is totally bounded, for every $\delta > 0$, there exist finite points $g_1, \cdots, g_{N(\delta)}$ such that

$$\mathfrak{M} \subset \bigcup_{i=1}^{N(\delta)} B(g_i, \delta),$$

where $B(g, \delta)$ is the open ball with center g and radius δ. Thus, for each $g \in \mathfrak{M}$, we can choose $\vartheta_\delta(g) \in \{g_1, \cdots, g_{N(\delta)}\}$ so that $||\vartheta_\delta(g) - g|| < \delta$. Then define

$$H_{n,\delta}(g) = H_n(\vartheta_\delta(g)), \quad H_\delta(g) = H(\vartheta_\delta(g)).$$

Note that $H_{n,\delta}$ and H_δ are approximations of H_n and H by (4.30), and $H_{n,\delta}$ and H_δ are indexed by finite set. Convergence of the finite dimensional distributions of H_n to those of H implies that

$$H_{n,\delta} \Rightarrow H_\delta \tag{4.31}$$

in $l^\infty(\mathfrak{M})$

Let $f : l^\infty(\mathfrak{M}) \to \mathbb{R}$ be bounded and continuous. Then it follows that

$$\begin{aligned}&|\mathbb{E}^* f(H_n) - \mathbb{E}f(H)| \\ &\leq |\mathbb{E}^* f(H_n) - \mathbb{E}f(H_{n,\delta})| + |\mathbb{E}f(H_{n,\delta}) - \mathbb{E}f(H_\delta)| \\ &\quad + |\mathbb{E}f(H_\delta) - \mathbb{E}f(H)|.\end{aligned}$$

From (4.31),

$$\limsup_{n\to\infty} |\mathbb{E}f(H_{n,\delta}) - \mathbb{E}f(H_\delta)| = 0.$$

Given $\varepsilon > 0$, let $K \subset l^\infty(\mathfrak{M})$ be a compact set such that

$$\mathbb{P}(H \in K^c) < \varepsilon,$$

and there exists $\tau > 0$, if $x \in K$, $||x - y||_{\mathfrak{M}} < \tau$,

$$|f(x) - f(y)| < \varepsilon,$$

and for δ small enough

$$\mathbb{P}(||H_\delta - H||_{\mathfrak{M}} \geq \tau) < \varepsilon.$$

Then

$$\begin{aligned}&|\mathbb{E}f(H_\delta) - \mathbb{E}f(H)| \\ &\leq C\mathbb{P}(\{H \in K^c\} \cup \{||H_\delta - H||_{\mathfrak{M}} \geq \tau\}) \\ &\quad + \sup\{|f(x) - f(y)| : x \in K, ||x - y||_{\mathfrak{M}} < \tau\} \\ &\leq C_1\varepsilon,\end{aligned}$$

which implies

$$\lim_{\delta\to 0} |\mathbb{E}f(H_\delta) - \mathbb{E}f(H)| = 0.$$

Furthermore, we have

$$\begin{aligned}&|\mathbb{E}f(H_n) - \mathbb{E}f(H_{n,\delta})| \\ &\leq C\mathbb{P}(H_{n,\delta} \in K^c_{\tau/2}) + \mathbb{P}(||H_n - H_{n,\delta}||_{\mathfrak{M}} \geq \tau/2) \\ &\quad + \sup\{|f(x) - f(y)| : x \in K, ||x - y||_{\mathfrak{M}} < \tau\},\end{aligned}$$

where $K_{\tau/2}$ is the $\tau/2$ open neighborhood of the set K for the sup norm. For big enough n, if $H_{n,\delta} \in K_{\tau/2}$ and $||H_{n,\delta} - H_n||_{\mathfrak{M}} \leq \tau/2$, then there exists $x \in K$ such that $||H_{n,\delta} - x||_{\mathfrak{M}} \leq \tau/2$ and $||x - H_n||_{\mathfrak{M}} \leq \tau$. Now, (4.30) implies that there is a δ_1 such that

$$\limsup_{n\to\infty} \mathbb{P}^*(||H_{n,\delta} - H_n||_{\mathfrak{M}} \geq \tau/2) < C\varepsilon$$

for all $\delta < \delta_1$ and (4.31) yields

$$\limsup_{n\to\infty} \mathbb{P}(H_{n,\delta} \in K^c_{\tau/2}) \leq \mathbb{P}(H_\delta \in K^c_{\tau/2}) \leq C\varepsilon.$$

Hence

$$\lim_{\delta\to 0} \limsup_{n\to\infty} |\mathbb{E}f(H_n) - \mathbb{E}f(H_{n,\delta})| = 0.$$

The proof is complete. □

Whether a given class $\mathfrak{M}$ is a P-Donsker class depends on the size of the class. In fact, a finite class of square integrable functions is always P-Donsker class, however, a class of all square integrable uniformly bounded functions is almost never a P-Donsker class. A relatively convenient way to measure the size of $\mathfrak{M}$ is to use the entropy. We give the definition of entropy in the following.

Definition 4.3. Let $(\mathfrak{M}, ||\cdot||)$ be a subset of a metric or pseudo-metric space. The *covering number* $N(\delta, \mathfrak{M}, ||\cdot||)$ is the minimal number of balls $\{h : ||h - f|| < \delta\}$ of radius δ needed to cover the set $\mathfrak{M}$ where f is the center of ball.

Note that the centers of the balls need not belong to $\mathfrak{M}$, but they should have finite norms.

Let $(\mathfrak{M}, ||\cdot||)$ be a subset of a metric or pseudo-metric space of real functions.

For two functions l and m with $l \leq m$, the bracket $[l, m]$ is the set of all functions f satisfying $l \leq f \leq m$, the $\delta-$bracket is the bracket $[l, m]$ with $||m - l|| < \delta$.

Definition 4.4. The *bracketing number* $N_{[]}(\delta, \mathfrak{M}, ||\cdot||)$ is the minimal number of $\delta-$brackets needed to cover the set $\mathfrak{M}$.

Definition 4.5. The *entropy* without bracketing is the logarithm of the covering number. The entropy with bracketing is the logarithm of the bracketing number.

In the study of weak convergence of empirical processes, we usually use the $L_r(\mathbb{Q})-$norm

$$||f||_{L_r(\mathbb{Q})} = (\int |f(x)|^r \mathbb{Q}(dx))^{1/r}, \qquad f \in \mathfrak{M}.$$

Definition 4.6. An *envelope function* of a class $\mathfrak{M}$ is any function $F(x)$ such that $|f(x)| \leq F(x)$ for every $x \in \mathbb{R}$ and $f \in \mathfrak{M}$.

In the following, we introduce some inequalities, and express these inequalities in abstract setting, using the Orlicz norm.

Let ψ be a nondecreasing convex function with $\psi(0) = 0$ and ξ a random variable. The *Orlicz norm* $||\xi||_\psi$ is defined as

$$||\xi||_\psi = \inf\{C > 0 : \mathbb{E}\psi(\frac{|\xi|}{C}) \leq 1\}.$$

The best-known example of the Orlicz norm is L_p-norm

$$||\xi||_p = (\mathbb{E}|\xi|^p)^{1/p},$$

which is the case of $\psi(x) = x^p$ for $p > 1$.

There is a quite important Orlicz norm, which is given by

$$\psi_p(x) = \exp(x^p) - 1$$

with $p \geq 1$. It is easy to obtain that

$$||\xi||_{\psi_p} \leq ||\xi||_{\psi_q} (\log 2)^{p/q}, \quad p \leq q,$$

$$||\xi||_{L_p} \leq p! ||\xi||_{\psi_1}, \quad p \geq 1.$$

Next, we will present some useful inequalities.

Lemma 4.7. *Let ξ be a random variable with*

$$\mathbb{P}(|\xi| > x) \leq C_1 \exp(-C_2 x^p)$$

for every x, where C_1 and C_2 are constants, and $p \geq 1$. Then

$$||\xi||_{\psi_p} \leq (\frac{1 + C_1}{C_2})^{1/p}.$$

Proof. By Fubini's theorem

$$\begin{aligned}
\mathbb{E}(\exp(D|\xi|^p) - 1) &= \mathbb{E} \int_0^{|\xi|^p} D \exp(Ds) ds \\
&= \int_0^{+\infty} \mathbb{P}(|\xi| > s^{1/p}) D \exp(Ds) ds \\
&\leq \int_0^{+\infty} C_1 D \exp((D - C_2)s) ds = \frac{C_1 D}{C_2 - D}.
\end{aligned}$$

$\frac{C_1 D}{C_2 - D}$ is less than or equal to 1 for $D^{-1/p}$ greater than or equal to $(\frac{1+C_1}{C_2})^{1/p}$ by the definition of the Orlicz norm. We obtain the lemma. □

Lemma 4.8. *Let ψ be a convex nondecreasing nonzero function with $\psi(0) = 0$ and*

$$\limsup_{x,y \to \infty} \frac{\psi(x)\psi(y)}{\psi(cxy)} < \infty$$

for some constant $c > 0$. Then for random variables $\xi_1, \cdots, \xi_m$,

$$|| \max_{1 \leq i \leq m} \xi_i ||_\psi \leq C \psi^{-1}(m) \max_{1 \leq i \leq m} ||\xi_i||_\psi$$

for a constant C depending only on ψ.

Proof. We first assume that $\psi(x)\psi(y) \leq \psi(cxy)$ for all $x, y \geq 1$. Then, for $y \geq 1$ and any $C > 0$,

$$\begin{aligned}
\max_{1 \leq i \leq m} \psi(\frac{|\xi_i|}{Cy}) &\leq \max_{1 \leq i \leq m} [\frac{\psi(c|\xi_i|/C)}{\psi(y)} + \psi(\frac{|\xi_i|}{Cy} 1_{\{\frac{|\xi_i|}{Cy} < 1\}}] \\
&\leq \sum_{i=1}^m \frac{\psi(c|\xi_i|/C)}{\psi(y)} + \psi(1).
\end{aligned}$$

Obviously,

$$\mathbb{E}\psi(\frac{|\xi_i|}{\max_{1\le i\le m}||\xi_i||_\psi}) \le \mathbb{E}\psi(\frac{|\xi_i|)}{||\xi_i||_\psi}) \le 1,$$

letting $C = c\max_{1\le i\le m}||\xi_i||_\psi$, we obtain

$$\mathbb{E}\psi(\frac{\max_{1\le i\le m}|\xi_i|}{Cy}) \le \max_{1\le i\le m}\psi(\frac{|\xi_i|}{cy}) \le \frac{m}{\psi(y)} + \psi(1).$$

When $\psi(1) \le 1/2$ and $y = \psi^{-1}(2m)$, we have $\mathbb{E}\psi(\frac{\max_{1\le i\le m}|\xi_i|}{Cy}) \le 1$, which is greater than 1 when $\psi(1) > 1/2$. Thus

$$||\max_{1\le i\le m}\xi_i||_\psi \le \psi^{-1}(2m)c\max_{1\le i\le m}||\xi_i||_\psi.$$

By the convexity of ψ and $\psi(0) = 0$, we have $\psi^{-1}(2m) \le 2\psi^{-1}(m)$, the conclusion is obtained.

For a general ψ, there are constants $a \le 1$ and $b > 0$ such that $g(x) = a\psi(bx)$ satisfies the conditions in the previous argument. Observe that

$$||\xi_i||_\psi \le \frac{||\xi_i||_g}{ab} \le \frac{||\xi_i||_\psi}{a},$$

we complete the proof. □

The previous inequality is useless for a supremum over infinitely many variables, for example, stochastic process. Such a case can be handled via a method known as chaining. To obtain similar inequality to stochastic process $\{X_t : t \in T\}$, we need to introduce several notions concerning the size of index set T. Let (T, d) be a metric or pseudo-metric space.

Lemma 4.9. *Let (T, d) be a pseudo-metric space, $\{X_t, t \in T\}$ a stochastic process indexed by T. Let ψ be a convex nondecreasing, nonzero function with $\psi(0) = 0$,*

$$\limsup_{x,y\to\infty}\frac{\psi^{-1}(xy)}{\psi^{-1}(x)\psi^{-1}(y)} < \infty, \qquad \limsup_{x\to\infty}\frac{\psi^{-1}(x^2)}{\psi^{-1}(x)} < \infty$$

and

$$||X_t - X_s||_\psi \le d(s,t), \quad s,t \in T.$$

Then for all finite subsets $S \subset T$, $t_0 \in T$ and $\varepsilon > 0$,

$$||\max_{t\in S}|X_t|||_\psi \le ||X_{t_0}||_\psi + C\int_0^D \psi^{-1}(N(x,T,d))dx, \tag{4.32}$$

$$||\max_{s,t\in S, d(s,t)<\varepsilon}|X_t - X_s|||_\psi \le C\int_0^{\varepsilon/2}\psi^{-1}(N(x,T,d))dx \tag{4.33}$$

for a constant C depending only on ψ, and D is the diameter of (T, d).

Proof. If (T, d) is not totally bounded, the right sides of (4.32) and (4.33) are infinite. Hence, we assume (T, d) is totally bounded and has diameter less than 1. For the sake of simplicity, we assume $t_0 \in S$ and $X_{t_0} = 0$. For any integer $k > 0$, let $\{s_1^k, \cdots, s_{N_k}^k\} \subset S$ be the center of $N_k \equiv N(2^{-k}, S, d)$ open balls of radius at most 2^{-k} and centers in S that cover S. Assume $S_0 = \{t_0\}$. Let $\pi_k : S \mapsto S_k$ be a map such that $d(s, \pi_k(s)) < 2^{-k}$ for all $s \in S$. There exists an integer k_S such that for $k \geq k_S$ and $s \in S$, $d(\pi_k(s), s) = 0$ since S is finite. Therefore, for $s \in S$

$$X_s = \sum_{k=1}^{k_S} (X_{\pi_k(s)} - X_{\pi_{k-1}(s)})$$

almost surely. Note that

$$d(\pi_{k-1}(s), \pi_k(s)) \leq d(s, \pi_k(s)) + d(s, \pi_{k-1}(s)) < 3 \cdot 2^{-k}.$$

Therefore,

$$\begin{aligned} || \max_{s \in S} |X_s| ||_\psi &\leq \sum_{k=1}^{k_S} || \max_{s \in S} |X_{\pi_k(s)} - X_{\pi_{k-1}(s)}| ||_\psi \\ &\leq 3C_\psi \sum_{k=1}^{k_S} 2^{-k} \psi^{-1}(N_k N_{k-1}) \\ &\leq C \sum_{k=1}^{k_S} 2^{-k} \psi^{-1}(N_k). \end{aligned}$$

This implies (4.32) since $N(2x, S, d) \leq N(x, T, d)$ for any $x > 0$, and then by bounding the sum in the last display by the integral in (4.32).

To prove (4.33), for $\varepsilon > 0$, let $V = \{(s, t) : s, t \in T, d(s, t) \leq \varepsilon\}$, and for $v := (t_v, s_v) \in V$ define

$$Y_v = X_{t_v} - X_{s_v}.$$

For $u, v \in V$, define the pseudo-metric

$$\rho(u, v) = ||Y_u - Y_v||_\psi.$$

We can assume that $\varepsilon \leq diam(T)$. Note that

$$diam_\rho(V) := \sup_{u, v \in V} \rho(u, v) \leq 2 \max_{v \in V} ||Y_v||_\psi \leq 2\varepsilon,$$

and furthermore,

$$\rho(u, v) \leq ||X_{t_v} - X_{t_u}||_\psi + ||X_{s_v} - X_{s_u}||_\psi \leq d(t_v, t_u) + d(s_v, s_u).$$

If $t_1, \cdots, t_N$ are the centers of a covering of T by open balls of radius at most x, then the set of open balls with centers in $\{(t_i, t_j) : 1 \leq i, j \leq N\}$ and $\rho-$radius $2x$ cover V. Furthermore, if the $2x$ ball about (t_i, t_j) has a non-empty intersection with V, then it is contained in a ball of radius $4x$ centered at a point in V. Thus

$$N(4x, V, \rho) \leq N^2(x, T, d).$$

Thus, applying (4.32) to the process Y, we find that

$$\begin{aligned} \| \max_{s,t \in S, d(s,t)<\varepsilon} |X_t - X_s| \|_\psi &\leq C \int_0^{2\varepsilon} \psi^{-1}(N(x, V, \rho))dx \\ &\leq C \int_0^{2\varepsilon} \psi^{-1}(N^2(x/4, T, d))dx \\ &\leq C' \int_0^{\varepsilon/2} \psi^{-1}(N(x, T, d))dx. \end{aligned}$$

□

Definition 4.7. A process $\{X_t : t \in T\}$ is separable if there is a countable set $T_0 \subset T$ and a subset $\Omega_0 \subset \Omega$ with $\mathbb{P}(\Omega_0) = 1$ such that for all $\omega \in \Omega_0$, $t \in T$, and $\varepsilon > 0$, $X(t, \omega)$ is in the closure of $\{X(s, \omega) : s \in T_0 \cap B(t, \varepsilon)\}$, where $B(t, \varepsilon)$ is $\varepsilon-$open ball of t.

If X is separable, it is easily seen that

$$\| \sup_{t \in T} |X_t| \|_\psi = \sup_{S \subset T,\ S \text{ finite}} \| \max_{t \in S} |X_t| \|_\psi$$

and similarly for $\| \sup_{s,t \in T,\ d(s,t)<\varepsilon} |X_t - X_s| \|_\psi$.

The following lemma is an easy consequence of the preceding lemma.

Lemma 4.10. *Let (T, d) be a pseudo-metric space, $\{X_t, t \in T\}$ a separable stochastic process. Let ψ be a convex nondecreasing nonzero function with $\psi(0) = 0$,*

$$\limsup_{x,y \to \infty} \frac{\psi^{-1}(xy)}{\psi^{-1}(x)\psi^{-1}(y)} < \infty, \qquad \limsup_{x \to \infty} \frac{\psi^{-1}(x^2)}{\psi^{-1}(x)} < \infty,$$

and

$$\|X_t - X_s\|_\psi \leq d(s, t), \quad s, t \in T.$$

Then for all $\varepsilon > 0$,

$$\| \max_{t \in T} |X_t| \|_\psi \leq \|X_{t_0}\|_\psi + C \int_0^D \psi^{-1}(N(x, T, d))dx, \tag{4.34}$$

$$\| \max_{s,t \in T, d(s,t)<\varepsilon} |X_t - X_s| \|_\psi \leq C \int_0^\varepsilon \psi^{-1}(N(x, T, d))dx \tag{4.35}$$

for a constant C depending only on ψ, and D is the diameter of (T, d).

If the stochastic process has multivariate normal finite-dimensional distributions (called as Gaussian process), we have the following lemma. In particular, we define a pseudo-metric

$$\rho_X^2(s, t) = \mathbb{E}[(X_t - X_s)^2],$$

where X is a Gaussian process.

Lemma 4.11. *Let (T, ρ_X) be a pseudo-metric space, $\{X_t, t \in T\}$ a separable Gaussian process indexed by T. Then for all $0 < \varepsilon < D$,*

$$||\max_{t\in T}|X_t|||_{\psi_2} \leq ||X_{t_0}||_{\psi_2} + C\int_0^D \sqrt{\log N(x, T, \rho_X)}dx, \tag{4.36}$$

$$||\max_{s,t\in T, \rho_X(s,t)<\varepsilon}|X_t - X_s|||_{\psi_2} \leq C\int_0^\varepsilon \sqrt{\log N(x, T, \rho_X)}dx \tag{4.37}$$

for a constant C depending only on ψ, and D is the diameter of (T, ρ_X).

Proof. It is easy to see that if $Z \backsim N(0,1)$,

$$\mathbb{E}\exp(\frac{Z^2}{c^2}) = \frac{1}{\sqrt{1-2/c^2}} < \infty$$

for $c^2 > 2$. Choosing $c = \sqrt{8/3}$, we have

$$\mathbb{E}\exp(\frac{Z^2}{c^2}) = 2.$$

Hence $||Z||_{\psi_2} = \sqrt{8/3}$. By homogeneity this yields $||aZ||_{\psi_2} = a\sqrt{8/3}$. Thus

$$||X_t - X_s||_{\psi_2} = \sqrt{8/3}\rho_X(s,t).$$

Furthermore,

$$\psi_2^{-1}(x) = \sqrt{\log(1+x)} \leq C\sqrt{\log x}, \quad x \geq 2.$$

Applying Lemma 4.10, we complete the proof. □

The previous lemma can be extended virtually without changing to *sub-Gaussian process*, a process X_t, $t \in T$ with respect to the pseudo-metric d on T satisfying

$$\mathbb{P}(|X_s - X_t| > x) \leq 2\exp(-\frac{x^2}{2d^2(s,t)}).$$

Example 4.1. Suppose that $\{\varepsilon_n\}_{n\geq 1}$ is a sequence of independent Rademacher random variables (that is, $\mathbb{P}(\varepsilon_i = \pm 1) = \frac{1}{2}$), and let

$$X_t = \sum_{i=1}^k t_i\varepsilon_i, \quad t = (t_1, t_2, \cdots, t_k) \in \mathbb{R}^k.$$

Then by Hoeffding's inequality

$$\mathbb{P}(|X_s - X_t| > x) \leq 2\exp(-\frac{x^2}{2||s-t||^2}),$$

i.e., $X_t, t \in \mathbb{R}^k$ is a sub-Gaussian process.

Next, we will introduce the symmetrization inequalities, which are the main tools to derive the weak convergence of functional index empirical processes.

To prove the weak convergence of

$$P_n(g) - P(g) = \frac{1}{n}\sum_{i=1}^{n}(g(X_i) - P(g)),$$

the main approaches is based on the principle of comparing the empirical processes to symmetrized processes

$$\mathbb{G}_n^0 g = \frac{1}{n}\sum_{i=1}^{n}\varepsilon_i g(X_i)$$

or

$$\mathbb{G}_n^+ g = \frac{1}{n}\sum_{i=1}^{n}\varepsilon_i(g(X_i) - P(g)),$$

where $\varepsilon_n, n \geq 1$, are i.i.d. Rademacher random variables. It will be shown that the law of large numbers or weak convergence holds for one of these processes if and only if it holds for the other two processes.

It will be convenient to generalize the inequality beyond the empirical process setting. We instead consider sums of independent stochastic processes $\{Z_i(g) : g \in \mathfrak{M}\}_{i\geq 1}$. The empirical process corresponds to taking $Z_i(g) = g(X_i) - Pg$. We do not assume any measurability of Z_i, but for computing outer expectations $\mathbb{E}^*$, it will be understood that the underlying probability space is a product space $(\mathbb{R}^n, \mathcal{B}^n, P^n) \times (\mathcal{Z}, \mathcal{C}, Q)$, where $(\mathbb{R}^n, \mathcal{B}^n, P^n)$ is the n times product space of $(\mathbb{R}, \mathcal{B}, P)$, $(\mathcal{Z}, \mathcal{C}, Q)$ is auxiliary probability space. Each Z_i is a function of the ith coordinate $(z, x) = (z_1, \cdots, z_n, x)$.

Lemma 4.12. *Let $Z_1, \cdots, Z_n$ be independent stochastic processes with mean 0. Then for any nondecreasing convex function $\Phi : \mathbb{R} \to \mathbb{R}$ and arbitrary function $S_i : \mathfrak{M} \to \mathbb{R}$,*

$$\mathbb{E}^*\Phi(\frac{1}{2}||\sum_{i=1}^{n}\varepsilon_i Z_i||_{\mathfrak{M}}) \leq \mathbb{E}^*\Phi(||\sum_{i=1}^{n} Z_i||_{\mathfrak{M}}) \leq \mathbb{E}^*\Phi(2||\sum_{i=1}^{n}\varepsilon_i(Z_i - S_i)||_{\mathfrak{M}}).$$

Proof. Let $Y_1, \cdots, Y_n$ be an independent copy of $Z_1, \cdots, Z_n$ defined on $(\mathbb{R}^n, \mathcal{B}^n, P^n) \times (\mathcal{Z}, \mathcal{C}, Q) \times (\mathbb{R}^n, \mathcal{B}^n, P^n)$ and depending on the last n coordinates. Since $\mathbb{E}Y_i = 0$, $\mathbb{E}^*\Phi(\frac{1}{2}||\sum_{i=1}^{n}\varepsilon_i Z_i||_{\mathfrak{M}})$ can be seen as an average of expressions of the type

$$\mathbb{E}_Z^*\Phi(\frac{1}{2}||\sum_{i=1}^{n} e_i(Z_i - \mathbb{E}Y_i)||_{\mathfrak{M}}),$$

where $(e_1, \cdots, e_n)$ ranges over $\{-1, 1\}^n$, $\mathbb{E}_Z^*$ is the outer expectation with respect to $Z_1, Z_2, \cdots, Z_n$ computed for given probability measure. The definitions of $\mathbb{E}_Y^*$, $\mathbb{E}_{Z,Y}^*$ are similar. By the convexity of Φ and the definition of the norm $||\cdot||_{\mathfrak{M}}$,

$$\mathbb{E}^*\Phi(\frac{1}{2}||\sum_{i=1}^{n} e_i(Z_i - \mathbb{E}Y_i)||_{\mathfrak{M}})$$

$$\leq \mathbb{E}^*_{Z,Y}\Phi(\frac{1}{2}||\sum_{i=1}^n e_i(Z_i - Y_i)||_{\mathfrak{M}}) = \mathbb{E}^*_{Z,Y}\Phi(\frac{1}{2}||\sum_{i=1}^n (Z_i - Y_i)||_{\mathfrak{M}})$$

by Jensen's inequality. Use the triangle inequality and convexity of Φ yields the left side inequality.

To prove the inequality on the right side, note that for fixed values of Z_i's we have

$$||\sum_{i=1}^n Z_i||_{\mathfrak{M}} = \sup_{f\in\mathfrak{M}} |\sum_{i=1}^n (Z_i(f) - \mathbb{E}Y_i(f))| \leq \mathbb{E}^*_Y \sup_{f\in\mathfrak{M}} |\sum_{i=1}^n (Z_i(f) - Y_i(f))|.$$

Since Φ is convex, Jensen's inequality yields

$$\Phi(||\sum_{i=1}^n Z_i||_{\mathfrak{M}}) \leq \mathbb{E}_Y\Phi(||\sum_{i=1}^n (Z_i(f) - Y_i(f))||^{*Y}_{\mathfrak{M}}),$$

where $*Y$ denotes the minimal measurable majorant of the supremum with respect to $Y_1, \cdots, Y_n$ with $Z_1, \cdots, Z_n$ fixed. Take outer expectation with respect to $Z_1, \cdots, Z_n$, we have

$$\mathbb{E}^*_Z\Phi(||\sum_{i=1}^n Z_i||_{\mathfrak{M}}) \leq \mathbb{E}^*_Z\mathbb{E}_Y\Phi(||\sum_{i=1}^n (Z_i(f) - Y_i(f))||_{\mathfrak{M}}).$$

Note that adding a minus sign in front of a term $(Z_i(f) - Y_i(f))$ has the effect of exchanging Z_i and Y_i. By construction of the underlying probability space

$$\mathbb{E}^*\Phi(||\sum_{i=1}^n e_i(Z_i - Y_i)||_{\mathfrak{M}})$$

is the same for any $n-$tuple $(e_1, \cdots, e_n)$ ranges over $\{-1, 1\}^n$, thus

$$\mathbb{E}^*\Phi(||\sum_{i=1}^n Z_i||_{\mathfrak{M}}) \leq \mathbb{E}^*_\varepsilon\mathbb{E}^*_{Z,Y}\Phi(||\sum_{i=1}^n \varepsilon_i(Z_i - Y_i)||_{\mathfrak{M}}).$$

Now, add and subtract S_i inside the right side and use the triangle inequality and convexity of Φ, we have

$$\mathbb{E}^*_\varepsilon\mathbb{E}^*_{Z,Y}\Phi(||\sum_{i=1}^n \varepsilon_i(Z_i - Y_i)||_{\mathfrak{M}})$$
$$\leq \frac{1}{2}\mathbb{E}^*_\varepsilon\mathbb{E}^*_{Z,Y}\Phi(2||\sum_{i=1}^n \varepsilon_i(Z_i - S_i)||_{\mathfrak{M}}) + \frac{1}{2}\mathbb{E}^*_\varepsilon\mathbb{E}^*_{Z,Y}\Phi(2||\sum_{i=1}^n \varepsilon_i(S_i - Y_i)||_{\mathfrak{M}}).$$

Finally, the repeated outer expectations $\mathbb{E}^*_\varepsilon\mathbb{E}^*_{Z,Y}$ can be replaced by a joint outer expectation $\mathbb{E}^* = \mathbb{E}^*_{\varepsilon,Z,Y}$. We complete the proof. □

By taking $Z_i(g) = g(X_i) - Pg$, we have the following corollary.

Corollary 4.2. *If Φ is a nondecreasing and convex function, then*

$$\mathbb{E}^*\Phi(\frac{1}{2}||\mathbb{G}^+_n||_{\mathfrak{M}}) \leq \mathbb{E}^*\Phi(||P_n - P||_{\mathfrak{M}}) \leq \mathbb{E}^*\Phi(2||\mathbb{G}^0_n||_{\mathfrak{M}}) \wedge \mathbb{E}^*\Phi(2||\mathbb{G}^+_n||_{\mathfrak{M}}).$$

4.3.2 Glivenko-Cantelli theorems

Although our interest is weak convergence, it is necessary to present the Glivenko-Cantelli theorems for completeness of the empirical process theory.

Theorem 4.6. *Let $\mathfrak{M}$ be a class of measurable functions such that*

$$N_{[]}(\varepsilon, \mathfrak{M}, L_1(P)) < \infty$$

for every $\varepsilon > 0$. Then $\mathfrak{M}$ is $P-$Glivenko-Cantelli, that is

$$||P_n - P||_{\mathfrak{M}} = \sup_{g \in \mathfrak{M}} |P_n g - Pg| \xrightarrow{a.s.} 0.$$

Proof. For any $\varepsilon > 0$, choose $N_{[]}(\varepsilon, \mathfrak{M}, L_1(P))$ $\varepsilon-$brackets $\{[l_i, u_i]\}$ whose union contains $\mathfrak{M}$ and satisfies $P(u_i - l_i) < \varepsilon$ for every i. We have

$$(P_n - P)g \leq (P_n - P)u_i + P(u_i - g) \leq (P_n - P)u_i + \varepsilon.$$

Similarly,

$$(P - P_n)g \leq (P - P_n)l_i + P(g - l_i) \leq (P - P_n)l_i + \varepsilon.$$

Thus

$$\sup_{g \in \mathfrak{M}} |(P_n - P)g| \leq \max_{1 \leq i \leq m} (P_n - P)u_i \vee \max_{1 \leq i \leq m} (P - P_n)l_i + \varepsilon$$

where the right side converges almost surely to ε by the strong law of large numbers. Then

$$\limsup_{n \to \infty} ||P_n - P||_{\mathfrak{M}} \leq \varepsilon$$

for every $\varepsilon > 0$. □

The next theorem is more simple.

Theorem 4.7. *Let $\mathfrak{M}$ be a class of $P-$measurable functions that is $L_1(P)-$bounded. Then $\mathfrak{M}$ is $P-$Glivenko-Cantelli if*

(1) $P^* F < \infty$;

(2)

$$\lim_{n \to \infty} \frac{\mathbb{E}^* \log N(\varepsilon, \mathfrak{M}_C, L_2(P_n))}{n} = 0$$

for all $C < \infty$ and $\varepsilon > 0$, where F is an envelope function of $\mathfrak{M}$, $\mathfrak{M}_C$ is the class of functions $\{g 1_{\{F \leq C\}} : g \in \mathfrak{M}\}$.

Proof. By Corollary 4.2, measurability of class $\mathfrak{M}$ and Fubini's theorem,

$$\begin{aligned}\mathbb{E}^*||P_n - P||_{\mathfrak{M}} &\leq 2\mathbb{E}_X \mathbb{E}_\varepsilon ||\frac{1}{n}\sum_{i=1}^n \varepsilon_i g(X_i)||_{\mathfrak{M}} \\ &\leq 2\mathbb{E}_X \mathbb{E}_\varepsilon ||\frac{1}{n}\sum_{i=1}^n \varepsilon_i g(X_i)||_{\mathfrak{M}_C} + 2\mathbb{P}^* F 1_{\{F > C\}}.\end{aligned}$$

For sufficiently large C, the last term is small enough. It suffices to show that the first term converges to 0 for fixed C. If $\mathfrak{G}$ is an $\varepsilon-$net over $\mathfrak{M}_C$ in $L_2(P_n)$, then it is also an $\varepsilon-$net over $\mathfrak{M}_C$ in $L_1(P_n)$. Thus

$$\mathbb{E}_\varepsilon||\frac{1}{n}\sum_{i=1}^n \varepsilon_i g(X_i)||_{\mathfrak{M}_C} \leq \mathbb{E}_\varepsilon||\frac{1}{n}\sum_{i=1}^n \varepsilon_i g(X_i)||_{\mathfrak{G}} + \varepsilon.$$

The cardinality of $\mathfrak{G}$ can be chosen equal to $N(\varepsilon, \mathfrak{M}_C, L_2(P_n))$. We use Lemma 4.8 with $\psi_2(x) = \exp(x^2) - 1$, thus the right side of the last display is bounded by a constant multiple of

$$\sqrt{1 + \log N(\varepsilon, \mathfrak{M}_C, L_2(P_n))} \sup_{g\in\mathfrak{G}} ||\frac{1}{n}\sum_{i=1}^n \varepsilon_i g(X_i)||_{\psi_2|X} + \varepsilon$$

where $||\cdot||_{\psi_2|X}$ is taken over $\varepsilon_1, \cdots, \varepsilon_n$ with $X_1, \cdots, X_n$. By Example 4.1 and Lemma 4.8,

$$||\frac{1}{n}\sum_{i=1}^n \varepsilon_i g(X_i)||_{\psi_2|X} \leq \sqrt{6}||\frac{1}{n}\sum_{i=1}^n g(X_i)||_{L_2}$$

$$\leq \sqrt{\frac{6}{n}}(P_n g^2)^{1/2} \leq \sqrt{\frac{6}{n}}C.$$

Thus

$$\sqrt{1 + \log N(\varepsilon, \mathfrak{M}_C, L_2(P_n))} \sup_{g\in\mathfrak{G}} ||\frac{1}{n}\sum_{i=1}^n \varepsilon_i g(X_i)||_{\psi_2|X} + \varepsilon \to \varepsilon$$

in outer probability. Then

$$\mathbb{E}_\varepsilon||\frac{1}{n}\sum_{i=1}^n \varepsilon_i g(X_i)||_{\mathfrak{M}_C} \to 0$$

in probability. This concludes the proof that

$$\mathbb{E}^*||P_n - P||_{\mathfrak{M}} \to 0.$$

Furthermore, $||P_n - P||_{\mathfrak{M}}^*$ is a reverse submartingale with respect to a suitable filtration, and hence almost sure convergence follows from the reverse submartingale convergence theorem. □

It is useful to specialize Theorem 4.7 to the case of indicator functions of some class of subsets $\mathfrak{C}$ of $\mathbb{R}$. Set

$$\Delta_n^{\mathfrak{C}}(x_1, \cdots, x_n) = \sharp\{C \bigcap \{x_1, \cdots, x_n\} : C \in \mathfrak{C}\}.$$

We use $\mathcal{C}$ to denote the indicator function class of $\mathfrak{C}$.

Theorem 4.8. *If $\mathfrak{C}$ is a $P-$measurable class of sets, and*

$$\lim_{n\to} \frac{\mathbb{E}\log \Delta_n^{\mathfrak{C}}(X_1, \cdots, X_n)}{n} = 0.$$

Then

$$||P_n - P||_{\mathcal{C}}^* \xrightarrow{a.s.} 0.$$

Proof. Note that for any $r > 0$,

$$N(\varepsilon, \mathcal{C}, L_r(P_n)) = N(\varepsilon^{r^{-1}\vee 1}, \mathcal{C}, L_\infty(P_n)) \le (\frac{2}{\varepsilon^{r^{-1}\vee 1}})^n,$$

and

$$||f-g||_{L_r(P_n)} = \{P_n|f-g|^r\}^{1/(r\vee 1)},$$

$$||f-g||_{L_\infty(P_n)} = \max_{1\le i\le n} |f(X_i) - g(X_1)|.$$

If we can prove

$$\Delta_n^{\mathfrak{C}}(X_1, \cdots, X_n) = N(\varepsilon, \mathcal{C}, L_\infty(P_n)) \tag{4.38}$$

we can obtain this theorem by Theorem 4.7.

Let $C_1, \cdots, C_k$ with $k = N(\varepsilon, \mathcal{C}, L_\infty(P_n))$ form an $\varepsilon-$net for $\mathfrak{C}$ for the $L_\infty(P_n))$ metric. If $C \in \mathfrak{C}$ satisfies

$$\max_{1\le i\le n}(1_{C\setminus C_j}(X_i) + 1_{C_j\setminus C}(X_i)) = \max_{1\le i\le n}|1_C(X_i) - 1_{C_j}(X_i)| < \varepsilon < 1$$

for some $j \in \{1, \cdots, k\}$, then the left side must be 0, and hence no X_i is in either $C\setminus C_j$ or $C_j\setminus C$. Thus

$$k = \sharp\{\{X_1, \cdots, X_n\}\bigcap C, C \in \mathfrak{C}\};$$

(4.38) is obtained. □

4.3.3 *Donsker theorem*

Now, we come to the central part of this section and will develop the Donsker theorem, or equivalently, uniform central limit theorem, for classes of functions and sets.

Here is the main result in this section.

Theorem 4.9. *Let $\{X_i\}_{i\ge 1}$ be a sequence of i.i.d. random variables defined on canonical space $(\mathbb{R}, \mathcal{B}, P)$, where P is the common distribution of $\{X_i\}_{i\ge 1}$. Suppose that $\mathfrak{M}$ is a class of measurable functions with envelope function F satisfying:*

(i)

$$\int_0^\infty \sup_Q \sqrt{\log N(x||F||_{Q,2}, \mathfrak{M}, L_2(Q))}dx < \infty$$

where the supremum is taken over all finitely discrete measures Q on $(\mathbb{R}, \mathcal{B})$ with $||F||_{Q,2}^2 = \int_{\mathbb{R}} F^2 dQ > 0$;

(ii)

$$P^*F^2 < \infty;$$

(iii) for all $\delta > 0$, the classes

$$\mathfrak{M}_\delta = \{f - g : f, g \in \mathfrak{M}, ||f-g||_{P,2} < \delta\}$$

and $\mathfrak{M}_\infty$ are $P-$measurable. Then $\mathfrak{M}$ is P-Donsker.

Proof. For fixed $f \in \mathfrak{M}$, the classical central limit theorems holds. From Theorem 4.5, it is enough to prove that $\mathfrak{M}$ is totally bounded in $L_2(P)$, and asymptotic equicontinuity for $\mathbb{G}_n$. Obviously, the latter is equivalent to the following statement: for every decreasing sequence $\delta \downarrow 0$,

$$||\mathbb{G}_n||_{\mathfrak{M}_\delta} \to 0$$

in outer probability measure. By Markov's inequality and Lemma 4.2,

$$P^*(||\mathbb{G}_n||_{\mathfrak{M}_\delta} > x) \leq \frac{2}{x} E^*(||\frac{1}{\sqrt{n}} \sum_{i=1}^n \varepsilon_i g(X_i)||_{\mathfrak{M}_\delta}).$$

By Hoeffding's inequality, the stochastic process

$$g \to \frac{1}{\sqrt{n}} \sum_{i=1}^n \varepsilon_i g(X_i)$$

is sub-Gaussian for the $L_2(P_n)-$seminorm

$$||g||_{L_2(P_n)} = \frac{1}{\sqrt{n}} \sqrt{\sum_{i=1}^n \varepsilon_i g^2(X_i)}.$$

By Lemma 4.11,

$$E^*(||\frac{1}{\sqrt{n}} \sum_{i=1}^n \varepsilon_i g(X_i)||_{\mathfrak{M}_\delta}) \leq C \int_0^\infty \sqrt{\log N(x, \mathfrak{M}_\delta L_2(P_n))} dx.$$

The set $\mathfrak{M}_\delta$ fits in a single ball of radius ε is larger than θ_n given by

$$\theta_n^2 = ||\frac{1}{n} \sum_{i=1}^n g^2(X_i)||_{\mathfrak{M}_\delta}.$$

Since the covering number of $\mathfrak{M}_\delta$ is bounded by covering number of $\mathfrak{M}_\infty$, and

$$N(\varepsilon, \mathfrak{M}_\infty, L_2(Q)) \leq N^2(\varepsilon, \mathfrak{M}, L_2(Q)),$$

we have

$$\begin{aligned}
E^*(||\frac{1}{\sqrt{n}} \sum_{i=1}^n \varepsilon_i g(X_i)||_{\mathfrak{M}_\delta}) &\leq \int_0^{\theta_n} \sqrt{\log N(x, \mathfrak{M}_\delta, L_2(P_n))} dx \\
&\leq \sqrt{2} ||F||_n \int_0^{\theta_n/||F||_n} \sqrt{\log N(x||F||_n, \mathfrak{M}, L_2(P_n))} dx \\
&\leq \sqrt{2} ||F||_n \int_0^{\theta_n/||F||_n} \sup_Q \sqrt{\log N(x||F||_{Q,2}, \mathfrak{M}, L_2(Q))} dx.
\end{aligned}$$

Furthermore,

$$\begin{aligned}
&E||F||_n \int_0^{\theta_n/||F||_n} \sup_Q \sqrt{\log N(x||F||_{Q,2}, \mathfrak{M}, L_2(Q))} dx \\
&\leq (E||F||_n^2)^{1/2} (E(\int_0^{\theta_n/||F||_n} \sup_Q \sqrt{\log N(x||F||_{Q,2}, \mathfrak{M}, L_2(Q))} dx)^2)^{1/2}.
\end{aligned}$$

If we can prove

$$\theta_n \le \delta + o_P(1)$$

in P^*, we can complete the proof. This will hold if

$$||P_n g^2 - Pg^2||_{\mathfrak{M}_\infty} \xrightarrow{P^*} 0 \qquad (4.39)$$

since

$$\sup\{Pg^2, g \in \mathfrak{M}_\delta\} \le \delta^2.$$

By the assumption, for any $f, g \in \mathfrak{M}_\infty$

$$P_n|f^2 - g^2| \le P_n(|f-g|(2F)) \le ||f-g||_{L_2(P_n)}||2F||_{L_2(P_n)}.$$

By the uniform entropy assumption (i), $N(\varepsilon||F||_n, \mathfrak{M}_\infty, L_2(P_n))$ is bounded by a fixed number, so its logarithm is certainly $o_{P_n}(n)$. By Theorem 4.7, (4.39) is true. Hence asymptotic equicontinuity holds.

It remains only to prove that $\mathfrak{M}$ is totally bounded in $L_2(P)$. By the previous arguments, there exists a sequence of discrete measure P_n with $||P_n g^2 - Pg^2||\mathfrak{M}_\infty$ converging to 0. Choose n sufficiently large so that the supremum is bounded by ε^2. By the assumption, $N(\varepsilon, \mathfrak{M}, L_2(P_n))$ is finite. But an $\varepsilon-$net for $\mathfrak{M}$ in $L_2(P_n)$ is $\sqrt{2}\varepsilon-$net for $\mathfrak{M}$ in $L_2(P)$, thus $\mathfrak{M}$ is totally bounded in $L_2(P)$. □

4.4 Weak convergence of empirical processes involving time-dependent data

In this section, we extend the result in section 4.3 to more general case. In (4.28), $\{X_i\}$ is a sequence of $\mathbb{R}-$vauled random variables. A more general version is replace the random variables by random elements taking values in some function space S, it means that empirical processes may involve the time evolution of the stochastic process X. The result in this section is from Kuelbs, Kurtz and Zinn (2013).

Let $(S, \mathcal{S}, P)$ be a probability space, and define $(\Omega, \mathcal{F}, \mathbb{P})$ to be the infinite product probability space $(S^N, \mathcal{S}^N, P^N)$ where $S^N = S \times S \times \cdots$. Let $X_i : \Omega \to S$ be the natural projections of Ω into the ith copy of S, and $\mathfrak{M}$ a subset of $L^2(S, \mathcal{S}, P)$ with

$$\sup_{f \in \mathfrak{M}} |f(s)| < \infty, \qquad s \in S.$$

Recall that $l^\infty(\mathfrak{M})$ is the set of the bounded real valued functions on $\mathfrak{M}$ with the sup-norm. We take i.i.d. copies $\{X_i\}_{i\ge1}$ of a process $\{X(t) : t \in [0,1]\}$. Consider stochastic process

$$\{\frac{1}{\sqrt{n}}\sum_{i=1}^{n}[1_{\{X_i(t)\le x\}} - \mathbb{P}(X(t) \le x)], x \in \mathbb{R}; t \in [0,1]\}$$

with time t. Our goal is to determine when these processes converge in distribution in some uniform sense to a mean zero Gaussian process.

Let F_t be the distribution function of $X(t)$. The modified distribution function $F(x, \lambda)$ is defined by

$$F_t(x, \lambda) = \mathbb{P}(X(t) < x) + \lambda \mathbb{P}(X(t) = x).$$

Let V be uniformly distribution random variable on $[0, 1]$,

$$\tilde{F}_t(x) = F_t(x, V)$$

be the distributional transform of F_t. *Rüschendorf (2009)* showed that $\tilde{F}_t(X(t))$ is uniform on $[0, 1]$. Firstly, a fundamental theorem is presented.

Theorem 4.10. *(Andersen (1988)) Let X be a random element with distribution P with respect to $\mathbb{P}$, and*

$$M = \sup_{f \in \mathfrak{M}} |f(X)|.$$

Let $Pf = \int f dP$, $||Pf||_{\mathfrak{M}} = \sup_{f \in \mathfrak{M}} |Pf| < \infty$ and assume that

(1)

$$\lim_{u \to \infty} u^2 \mathbb{P}^*(M > u) = 0;$$

(2) $\mathfrak{M}$ is $P-$pre-Gaussian;

(3) there exists a centered Gaussian process $\{G(f) : f \in \mathfrak{M}\}$ with L_2 distance d_G such that G is sample bounded and uniformly d_G continuous on $\mathfrak{M}$, and for some $K > 0$, all $f \in \mathfrak{M}$ and all $\varepsilon > 0$,

$$\sup_{u>0} u^2 \mathbb{P}^*(\sup_{g: d_G(g,f)<\varepsilon} |f - g| > u) \leq K\varepsilon^2$$

Then $\mathfrak{M}$ is a $P-$Donsker class.

Theorem 4.11. *Let*

$$\rho^2(s, t) = \mathbb{E}(H(s) - H(t))^2,$$

for some centered Gaussian process H that is sample bounded and uniformly continuous on $([0, 1], \rho)$ with probability one. Furthermore, assume that for some $L < \infty$ and all $\varepsilon > 0$,

$$\sup_{t \in [0,1]} \mathbb{P}^*(\sup_{\{s: \rho(s,t) \leq \varepsilon\}} |\tilde{F}_t(X(s)) - \tilde{F}_t(X(t))| > \varepsilon^2) \leq L\varepsilon^2, \quad (4.40)$$

and $\mathfrak{D}([0, 1])$ is a collection of real valued functions on $[0, 1]$ such that $\mathbb{P}(X(\cdot) \in \mathfrak{D}([0, 1])) = 1$. Let

$$\mathcal{D} = \{C_{s,x} : s \in [0, 1], X \in \mathbb{R}\},$$

where

$$D_{s,x} = \{z \in \mathfrak{D}([0, 1]) : z(s) \leq x\}$$

for $s \in [0, 1], x \in \mathbb{R}$. Then weak convergence of

$$\{\frac{1}{\sqrt{n}} \sum_{i=1}^{n} [1_{\{X_i(t) \in D\}} - \mathbb{P}(X(t) \in D)] : t \in [0, 1], D \in \mathcal{D}\}$$

in $l^\infty(\mathcal{D})$ holds.

Remark 4.1. The relationship between $\{X(t)\}$ and $\rho(s,t)$ is given in (4.40), which allows one to establish the limiting Gaussian process.

To prove theorem, we need some lemmas.

Lemma 4.13. *Assume that for some $L < \infty$ and all $\varepsilon > 0$,*

$$\sup_{\{s,t\in[0,1]:\rho(s,t)\leq\varepsilon\}} \mathbb{P}^*(|\tilde{F}_t(X(s)) - \tilde{F}_t(X(t))| > \varepsilon^2) \leq L\varepsilon^2. \tag{4.41}$$

Then for all $x \in \mathbb{R}$,

$$\mathbb{P}(X(s) \leq x < X(t)) \leq (L+1)\rho^2(s,t) \tag{4.42}$$

and by symmetry,

$$\begin{aligned}\mathbb{E}|1_{\{X(t)\leq x\}} - 1_{\{X(s)\leq x\}}| &= \mathbb{P}(X_t \leq x < X_s) + \mathbb{P}(X_s \leq x < X_t)\\ &\leq 2(L+1)\rho^2(s,t).\end{aligned}$$

Further, we have

$$\sup_{x\in\mathbb{R}} |F_t(x) - F_s(x)| \leq 2(L+1)\rho^2(s,t).$$

Proof. Since $\tilde{F}_t$ is nondecreasing and $x \leq y$ implies $F_t(x) \leq \tilde{F}_t(y)$, we have

$$\begin{aligned}&\mathbb{P}(X(s) \leq x < X(t))\\ &\leq \mathbb{P}(F_t(x) \leq \tilde{F}_t(X(t)), \tilde{F}_t(X(s)) \leq \tilde{F}_t(x))\\ &\leq \mathbb{P}(F_t(x) \leq \tilde{F}_t(X(t)) \leq F_t(x) + \rho^2(s,t), \tilde{F}_t(X(s)) \leq \tilde{F}_t(x))\\ &\quad + \mathbb{P}(\tilde{F}_t(X(t)) > F_t(x)) + \rho^2(s,t), \tilde{F}_t(X(s)) \leq \tilde{F}_t(x))\end{aligned}$$

and hence

$$\begin{aligned}&\mathbb{P}(X(s) \leq x < X(t))\\ &\leq \mathbb{P}(F_t(x) \leq \tilde{F}_t(X(t)) \leq F_t(x) + \rho^2(s,t))\\ &\quad + \mathbb{P}(|\tilde{F}_t(X(t)) - \tilde{F}_t(X(s))| > \rho^2(s,t)).\end{aligned}$$

Now (4.41) implies for all $s, t \in [0,1]$ that

$$\mathbb{P}(|\tilde{F}_t(X(t)) - \tilde{F}_t(X(s))| > \rho^2(s,t)) \leq L\rho^2(s,t).$$

Therefore,

$$\mathbb{P}(X(s) \leq x < X(t)) \leq \rho^2(s,t) + L\rho^2(s,t)$$

since $\tilde{F}_t(X(t))$ is uniform on $[0,1]$, i.e., (4.42) holds.

The last conclusion follows by moving the absolute values outside the expectation. □

Lemma 4.14. *Assume that (4.41) holds. Let*

$$\tau((s,x),(t,y)) = [\mathbb{E}(1_{\{X(s)\le x\}} - 1_{\{X(t)\le y\}})^2]^{1/2},$$

$$\mathcal{G} = \{1_{\{X(s)\le x\}} : s \in [0,1], x \in \mathbb{R}\}.$$

Then

$$\tau^2((s,t),(t,y)) \le \min_{u\in\{s,t\}} |F_u(y) - F_u(x)| + (2L+2)\rho^2(t,s). \tag{4.43}$$

Moreover, if Q denotes the rational numbers, there is a countable dense set E_0 of $([0,1],\rho)$ such that $\mathcal{G}_0 = \{1_{\{X(s)\le x\}} : (s,x) \in E_0 \times Q\}$ is dense in $(\mathcal{G},\tau)$.

Proof.

$$\begin{aligned}
&\tau^2((s,x),(t,y))\\
&= \mathbb{E}|1_{\{X(s)\le x\}} - 1_{\{X(t)\le y\}}| \le \mathbb{E}|1_{\{X(t)\le y\}} - 1_{\{X(t)\le x\}}| + \mathbb{E}|1_{\{X(t)\le x\}} - 1_{\{X(s)\le x\}}|\\
&= |F_t(y) - F_t(x)| + \mathbb{P}(X(s) \le x < X(t)) + \mathbb{P}(X(t) \le x < X(s))\\
&\le |F_t(y) - F_t(x)| + (2+2L)\rho^2(s,t)
\end{aligned}$$

by using the symmetry in s and t and (4.42).

Similarly,

$$\tau^2((s,x),(t,y)) \le |F_s(y) - F_s(x)| + (2+2L)\rho^2(s,t).$$

Combining above two inequalities for τ, the proof of (4.43) holds. Since $([0,1],\rho)$ is totally bounded, there is a countable dense set E_0 of $([0,1],\rho)$, and the proof is completed by the right continuity of the distribution functions and (4.43). □

Lemma 4.15. *Assume that (s,x) and (t,y) satisfy*

$$\tau((s,x),(t,y)) = ||1_{\{X(s)\le x\}} - 1_{\{X(t)\le y\}}||_2 \le \varepsilon,$$

$\rho(s,t) \le \varepsilon$, and (4.41) holds. Then, for $c = (2L+2)^{1/2} + 1$,

$$|F_t(x) - F_t(y)| \le (c\varepsilon)^2.$$

Proof. In fact,

$$\begin{aligned}
&|F_t(y) - F_t(x)|^{1/2}\\
&= ||1_{\{X(t)\le y\}} - 1_{\{X(t)\le x\}}||_2 \le ||1_{\{X(s)\le x\}} - 1_{\{X(t)\le y\}}||_2\\
&\quad + ||1_{\{X(s)\le x\}} - 1_{\{X(t)\le x\}}||_2\\
&\le \varepsilon + (\mathbb{P}(X(s) \le x < X(t)) + \mathbb{P}(X(t) \le x < X(s)))^{1/2}\\
&\le \varepsilon + (2L\varepsilon^2 + 2\varepsilon^2)^{1/2}\\
&= [(2L+2)^{1/2} + 1]\varepsilon = c\varepsilon.
\end{aligned}$$

The lemma is proved. □

Lemma 4.16. *If (4.40) holds, $c = (2L+2)^{1/2} + 1$, and*

$$\lambda((s,x),(t,y)) = \max\{\tau((s,x),(t,y)), \rho(s,t)\},$$

then for all (t,y) and $\varepsilon > 0$,

$$\mathbb{P}^*(\sup_{\{(s,x):\lambda((t,y),(s,x))\le\varepsilon\}} |1_{\{X(t)\le y\}} - 1_{\{X(s)\le x\}}| > 0) \le 2(c^2 + L + 1)\varepsilon^2.$$

Proof. Firstly,

$$\begin{aligned}
&\mathbb{P}^*(\sup_{\{(s,x):\lambda((t,y),(s,x))\le\varepsilon\}} |1_{\{X(t)\le y\}} - 1_{\{X(s)\le x\}}| > 0) \\
&= \mathbb{P}^*(\sup_{\{(s,x):\lambda((t,y),(s,x))\le\varepsilon\}} 1_{\{X(t)\le y, X(s)>x\}} + 1_{\{X(s)\le x, X(t)>y\}} > 0) \\
&\le \mathbb{P}^*(\sup_{\{(s,x):\lambda((t,y),(s,x))\le\varepsilon\}} 1_{\{\tilde{F}_t(X(t))\le \tilde{F}_t(y), F_t(x)\le \tilde{F}_t(X(s))\}} > 0) \\
&\quad + \mathbb{P}^*(\sup_{\{(s,x):\lambda((t,y),(s,x))\le\varepsilon\}} 1_{\{\tilde{F}_t(X(s))\le \tilde{F}_t(x), F_t(y)\le \tilde{F}_t(X(t))\}} > 0) \\
&=: I + II,
\end{aligned}$$

by using the fact that $x < y$ implies $F_t(x) \le \tilde{F}_t(y)$. By using Lemma 4.15 and (4.40),

$$\begin{aligned}
I &\le \mathbb{P}^*(\sup_{\{(s,x):\lambda((t,y),(s,x))\le\varepsilon\}} 1_{\{\tilde{F}_t(X(t))\le F_t(y), F_t(y)-(c\varepsilon)^2\le \tilde{F}_t(X(s))\}} > 0) \\
&\le \mathbb{P}^*(\sup_{\{(s,x):\lambda((t,y),(s,x))\le\varepsilon\}} 1_{\{\tilde{F}_t(X(t))\le F_t(y), F_t(y)-(c\varepsilon)^2\le \tilde{F}_t(X(t))+\varepsilon^2\}} > 0) + L\varepsilon^2 \\
&\le \mathbb{P}(F_t(y) - (c\varepsilon)^2 - \varepsilon^2 \le \tilde{F}_t((X_t)) \le F_t(y) + L\varepsilon^2) \\
&= (c^2 + L + 1)\varepsilon^2.
\end{aligned}$$

For II, since $\tilde{F}_t(x) \le F_t(x)$ for all x, Lemma 4.15, and the definition of L, we have

$$\begin{aligned}
II &\le \mathbb{P}(\tilde{F}_t(X(t)) - \varepsilon^2 \le F_t(y) + (c\varepsilon)^2, F_t(y) \le \tilde{F}_t(X(t))) + L\varepsilon^2 \\
&\le (c^2 + L + 1)\varepsilon^2.
\end{aligned}$$

Combining the estimates for I and II obtains the lemma. □

Lemma 4.17. *Let E_0 be a countable dense subset of $([0,1], \rho)$. Then there exits a sequence of partitions $\{\mathcal{A}_n : n \ge 0\}$ of $E_0 \times \mathbb{R}$ such that*

$$\lim_{r\to\infty} \sup_{(t,y)\in E_0\times\mathbb{R}} \sum_{n\ge r} 2^{n/2} \Delta_\tau(A_n((t,y))) = 0 \tag{4.44}$$

and

$$Card(\mathcal{A}_n) \le 2^{2^n},$$

where $\Delta_\tau(A)$ is the diameter of A with respect to τ.

The proof of this lemma is with strong technical skills, we omit it here. The readers can find details in *Kuelbs, Kurtz and Zinn (2013)*.

The proof of Theorem 4.11.

To prove this theorem, we verify the conditions in Theorem 4.10.

Let Q denote the set of rational numbers. Then, if we restrict the partitions $\mathcal{A}_n$ of $E_0 \times \mathbb{R}$ in Lemma 4.17 to $E_0 \times Q$, then

$$\lim_{r\to\infty} \sup_{(t,y)\in E_0\times Q} \sum_{n\geq r} 2^{n/2}\Delta_\tau(A_n((t,y))) = 0, \tag{4.45}$$

and $(E_0 \times Q, \tau)$ is totally bounded.

Let $\{G_{(s,x)} : (s,x) \in E \times \mathbb{R}\}$ be a centered Gaussian process with

$$\mathbb{E}(G_{(s,x)}G_{(t,y)}) = \mathbb{P}(X(s) \leq x, X(t) \leq y).$$

Then, G has L_2 distance τ, it is uniformly continuous on $(E_0 \times Q, \tau)$ by (4.45) and the Talagrand continuity theorem (Theorem 1.4.1, Talagrand (2005)). Hence if $\{H_{(s,x)} : (s,x) \in E_0 \times Q\}$ is a centered Gaussian process with

$$\mathbb{E}(H_{(s,x)}H_{(t,y)}) = \mathbb{P}(X(s) \leq x, X(t) \leq y) - \mathbb{P}(X(s) \leq x)\mathbb{P}(X(t) \leq y),$$

then

$$\mathbb{E}((H_{(s,x)} - H_{(t,y)})^2) = \tau^2((s,x),(t,y)) - (\mathbb{P}(X(s) \leq x) - \mathbb{P}(X(t) \leq y))^2.$$

Hence the L_2 distance of H is smaller than that of G, and therefore the process H is uniformly continuous on $(E_0 \times Q, d_H)$, where

$$d_H((s,x),(t,y)) = \rho(1_{\{X(s)\leq x\}}, 1_{\{X(t)\leq y\}}).$$

By Lemma 4.14 the set $E_0 \times Q$ is dense in $([0,1] \times \mathbb{R}, \tau)$, and

$$d_H((s,x),(t,y)) \leq \tau((s,x),(t,y)),$$

we also have that $E_0 \times Q$ is dense in $([0,1] \times \mathbb{R}, d_H)$.

Thus the Gaussian process $\{H_{(s,x)} : (s,x) \in [0,1] \times \mathbb{R}\}$ has a uniformly continuous version, which we also denote by H, and since (E, d_H) is totally bounded, the sample functions are bounded on $[0,1]$ with probability one.

By the definition of $d_H((s,x),(t,y))$, the continuity of H on $([0,1] \times \mathbb{R}, d_H)$ implies condition (2) in Theorem 4.10 is satisfied. Condition (1) in Theorem 4.10 is also satisfied, since $1_{\{X(t)\leq y\}}$ is bounded. Thus, it is enough to verify condition (3) of Theorem 4.10.

Let $\mathcal{H} = \{I_{\{X_s\leq x\}} : (s,x) \in E \times \mathbb{R}\}$. Since $f = 1_{\{X(s)\leq x\}} \in \mathcal{H}$ with the point $(s,x) \in [0,1] \times \mathbb{R}$, for the centered Gaussian process

$$\{G_f : f \in \mathfrak{M}\}$$

in (3) of Theorem 4.10, for $(s,x) \in [0,1] \times \mathbb{R}$, we take the process

$$\tilde{G}_{(s,x)} = G_{(s,x)} + \tilde{H}_s$$

where $\{\tilde{H}_s : s \in [0,1]\}$ is a Gaussian process whose law is that of the process $\{H_s : s \in [0.1]\}$ given in the theorem. Furthermore, $\{G_{(s,x)} : (s,x) \in [0,1] \times \mathbb{R}\}$ is a uniformly continuous and sample bounded version of the Gaussian process, also denoted by $G_{(s,x)}$, but defined above on $E_0 \times Q$. Since $E_0 \times Q$ is dense in $(E \times \mathbb{R}, \tau)$, the extension to all of $[0,1] \times \mathbb{R}$ again follows.

Therefore, $\tilde{G}$ is sample bounded and uniformly continuous on $[0,1] \times \mathbb{R}$ with respect to its L_2 distance

$$d_{\tilde{G}}((s,x),(t,y)) = \{\tau^2((s,x),(t,y)) + \rho^2(s,t)\}^{1/2}.$$

We have

$$\{(s,x) : \lambda((s,x),(t,y)) \leq \epsilon\} \supseteq \{(s,x) : d_{\tilde{G}}((s,x),(t,y)) \leq \varepsilon\}.$$

For a random variable Z bounded by one,

$$\sup_{t>0} t^2 \mathbb{P}(|Z| > t) \leq \mathbb{P}(|Z| > 0),$$

thus

$$\sup_{u>0} u^2 \mathbb{P}^*(\sup_{g:d_H(g,f)<\varepsilon} |f-g| > u) \leq \mathbb{P}^*(\sup_{g:d_H(g,f)<\varepsilon} |f-g| > 0).$$

Condition (3) of Theorem 4.10 now follows by Lemma 4.15.

4.5 Two examples of applications in statistics

Study of empirical processes is an important subject in statistics. Here, we just present two simple examples to show the importance of weak convergence of empirical processes, when people need to obtain the limiting distributions of some statistics.

4.5.1 *Rank tests in complex survey samples*

Data from complex survey samples such as national health and nutrition examination survey series in the US or the British household panel survey in the UK are increasingly being used for statistical analyses. Observations in these surveys are sampled with known unequal sampling distributions, and the sampling is typically correlated.

Data analysis in complex survey samples is quite different from the independently-sampled data analysis. The classical derivation of asymptotic distribution for rank tests is based strongly on exchangeability, which is not present in complex survey samples. Thus, people consider the test statistic as a Hadamard-differentiable functional of the estimated cumulative distribution function in each group, then obtain the asymptotic distribution for rank tests through the central limit theorem of empirical process. This example is from *Lumley (2013)*.

In a sequence of N_k populations and n_k samples, a real valued variables Y is observed together with a binary group G that divides the population into groups of size M_0 and M_1.

A sample, s, of n units is draw from the finite population using some probability sampling method with sampling probabilities π_i, and corresponding sampling weights π_i^{-1}, and the values of Y_i and G_i is observed for the sampled units. The estimated population ranks $\hat{R}_i$ are defined by

$$\hat{R}_i = \hat{F}_n(Y_i),$$

$$\hat{F}_n(y) = \frac{1}{\hat{N}} \sum_{j \in s} \frac{1}{\pi_j} 1_{\{Y_j \le y\}},$$

where $\hat{N} = \sum_{j \in s} \frac{1}{\pi_j}$. A rank test statistic is then defined by

$$\hat{T}_n = \frac{1}{\hat{M}_0} \sum_{i \in s_0} \frac{1}{\pi_i} g(\hat{R}_i) - \frac{1}{\hat{M}_1} \sum_{i \in s_1} \frac{1}{\pi_i} g(\hat{R}_i),$$

where g is a given function, $s_l = \{i \in s, G_i = l\}$, $\hat{M}_l = \sum_{j \in s_l} \frac{1}{\pi_j}$, $l = 0$ or 1.

Now, write

$$D_Y(y) = F_{0Y}(y) - F_{1Y}(y),$$

$$\hat{D}_n(y) = \hat{F}_{0n}(y) - \hat{F}_{1n}(y),$$

where F_{lY} denotes the conditional distribution function of Y given $G = l$ in the superpopulation, and

$$\hat{F}_{ln}(y) = \frac{1}{\hat{M}_l} \sum_{i \in s} \frac{1}{\pi_i} 1_{\{Y_i \le y\}}.$$

Theorem 4.12. *(Lumley (2013)) Assume that g is a continuous differentiable function, and the marginal distribution of Y is absolutely continuous with finite forth moment. Let*

$$\delta_Y = \int_{\mathbb{R}} g(y) dD_Y(y).$$

Then $\sqrt{n}(\hat{T}_n - \delta_Y)$ is asymptotically normal.

Proof. The finiteness of fourth moment of Y implies that

$$\sqrt{n}(\hat{D}_n - D_Y, \hat{F}_n - F_Y)$$

converges weakly to a Gaussian process $Z_\pi = (Z_{\pi 1}, Z_{\pi 2})$ by Theorem 4.9. The details can be found in Theorem 3 in Lumley (2013). We can write

$$\hat{T}_n = \int_{\mathbb{R}} g(\hat{F}_n) d\hat{D}_n.$$

As g is differentiable and bounded, $\phi(D, F) = \int g(F) dD$ is Hadamard differentiable with Hadamard derivative

$$\phi'_{D,F}(\alpha, \beta) \int g(F) d\alpha + \int \beta g'(F) dD.$$

Then

$$\sqrt{n}(\hat{T}_n - \delta_Y) \xrightarrow{d} \int g dZ_{\pi 1} + \int Z_{\pi 2} g' dD_Y,$$

from the Donsker Theorem and the functional delta method. The proof is complete. □

4.5.2 *M-estimator*

M-estimators stands for a broad class of estimators, which are obtained as the minima of functions of the data. LSE is important examples of M-estimators. More generally, an M-estimator may be defined to be a zero of an estimating function, the derivative of another statistical function.

Let $X_1, X_2, \cdots$ be i.i.d. copies of a random variable X taking values in $\mathbb{R}$ and with distribution P.

Let Θ be a parameter space and for $\theta \in \Theta$, $\gamma_\theta : \mathbb{R} \to \mathbb{R}$ some loss function. We assume $P(|\gamma_\theta|) < \infty$ for all $\theta \in \Theta$. We estimate the unknown parameter

$$\theta_0 := \arg\min_{\theta\in\Theta} P(\gamma_\theta).$$

By the $M-$estimator

$$\theta_n := \arg\min_{\theta\in\Theta} P_n(\gamma_\theta),$$

where

$$P_n = \frac{1}{n}\sum_{i=1}^{n} \delta_{X_i}.$$

We assume that θ_0 exists and is unique. Our interest is to obtain the asymptotic normality of estimator $\hat{\theta}_n$. Firstly, three conditions are presented.

Condition a. There is a function $\psi_0 : \mathbb{R} \to \mathbb{R}$, with $P(\psi_0^2) < \infty$ such that

$$\lim_{\theta\to\theta_0} \frac{||\gamma_\theta - \gamma_{\theta_0} - \psi_0(\theta - \theta_0)||_{2,P}}{|\theta - \theta_0|} = 0.$$

Condition b.

$$P(\gamma_\theta) - P(\gamma_{\theta_0}) = \frac{1}{2}C(\theta - \theta_0)^2 + o(|\theta - \theta_0|) + o(|\theta - \theta_0|^2)$$

as $\theta \to \theta_0$, where $C > 0$.

Condition c. There exists an $\varepsilon > 0$ such that the class

$$\{g_\theta : |\theta - \theta_0| < \varepsilon\}$$

is $P-$Donsker class with envelope G with $P(G^2) < \infty$, where

$$g_\theta = \frac{\gamma_\theta - \gamma_{\theta_0}}{|\theta - \theta_0|}.$$

Theorem 4.13. *Suppose conditions a, b and c are satisfied. Then*

$$\sqrt{n}(\hat{\theta}_n - \theta_0) \xrightarrow{d} N(0, \frac{J}{C^2}),$$

where $J = \int \psi_0^2 dP$.

Proof. We may write

$$\begin{aligned}
0 \geq P_n(\gamma_{\hat{\theta}_n} - \gamma_{\theta_0}) &= (P_n - P)(\gamma_{\hat{\theta}_n} - \gamma_{\theta_0}) + P(\gamma_{\hat{\theta}_n} - \gamma_{\theta_0}) \\
&= (P_n - P)(g_\theta)|\theta - \theta_0| + P(\gamma_{\hat{\theta}_n} - \gamma_{\theta_0}) \\
&= (P_n - P)\psi_0(\hat{\theta}_n - \theta_0) + o_{\mathbb{P}}(n^{-1/2}) + P(\gamma_{\hat{\theta}_n} - \gamma_{\theta_0}) \\
&= (P_n - P)\psi_0(\hat{\theta}_n - \theta_0) + o_{\mathbb{P}}(n^{-1/2})|\hat{\theta}_n - \theta_0| \\
&\quad + \frac{1}{2}C(\theta - \theta_0)^2 + o(|\hat{\theta}_n - \theta_0|^2).
\end{aligned}$$

This implies

$$|\hat{\theta}_n - \theta_0| = O_{\mathbb{P}}(n^{-1/2}).$$

Then

$$|C^{1/2}(\hat{\theta}_n - \theta_0) + C^{-1/2}(P_n - P)\psi_0 + o_{\mathbb{P}}(n^{-1/2})|^2 = o_{\mathbb{P}}(n^{-1}).$$

Therefore,

$$\hat{\theta}_n - \theta_0 = -C^{-1}(P_n - P)\psi_0 + o_{\mathbb{P}}(n^{-1/2}).$$

The result follows by condition c and $P(\psi_0) = 0$. □

Bibliography

Aalen, O. (1977). Weak convergence of stochastic integrals related to counting processes. *Probability Theory and Related Fields* **38**, 261-277.
Aldous, D. (1978). Stopping times and tightness. *The Annals of Probability* **6**, 335-340.
Andersen, N., Giné, E., Ossiander, M., Zinn, J. (1988). The central limit theorem and the law of iterated logarithm for empirical processes under local conditions. *Probability Theory and Related Fields* **77**, 271-305.
Aldous, D., Eagleson, G. (1978). On mixing and stability of limit theorems. *The Annals of Probability* **6**, 325-331.
Applebaum (2009). *Lévy Processes and Stochastic Calculus*, Cambridge Studies in Advanced Mathematics.
Avram, F., Taqqu, M. (1992). Weak convergence of sums of moving averages in the α-stable domain of attraction. *The Annals of Probability* **20**, 483-503.
Basrak, B., Segers, G. (2009). Regularly varying multivariate time series. *Stochastic Processes and their Applications* **119**, 1055-1080.
Basrak, B., Krizmanić, D., Segers, G. (2012). A functional limit theorem for dependent sequences with infinite variance stable limits. *The Annals of Probability* **40**, 2008-2033.
Billingsley, P. (1999). *Convergence of Probability Measures*, John Wiley and Sons.
Can, S., Mikosch, T., Samorodnitsky, G. (2010). Weak convergence of the function-indexed integrated periodogram for infinite variance processes. *Bernoulli* **16**, 995-1015.
Caner, M. (1997). Weak convergence to a matrix stochastic integral with stable processes. *Econometric Theory* **13**, 506-528.
Chan, N., Zhang, R. (2012). Marked empirical processes for non-stationary time series. Forthcoming in *Bernoulli*.
Chatterji, S. (1974). A principle of subsequences in probability theory: the central limit theorem. *Advances in Mathematics* **13**, 31-54.
Davis, R., Resnick, S. (1986). Limit theory for the sample covariance and correlation functions of moving averages. *The Annals of Statistics* **14**, 533-538.
Davis, R., Mikosch, T. (1998). The sample autocorrelations of heavy-tailed processes with applications to ARCH. *The Annals of Statistics* **26**, 2049-2080.
Dellacherie, C., Meyer, P. (1982). *Probabilities and Potential, volume B*, Amsterdam.
Dette, H., Podolskij, M. (2008). Testing the parametric form of the volatility in continuous time diffusion modelsa stochastic process approach. *Journal of Econometrics* **143**, 56-73.
Dudley, R. (1978). Central limit theorems for empirical measures. *The Annals of Probability* **143**, 899-929.
Ethier, S., Kurtz, T. (1986). *Markov Processes: Characterization and Convergence*, John Wiley and Sons.

Feller, W. (1971). *An Introduction to Probability and its Applications*, John Wiley and Sons.
Fermanian, J., Radulovic, D., Wegkamp, M. (2004). Weak convergence of empirical copula processes. *Bernoulli* **10**, 847-860.
Hall, P. (1977). Martingale invariance principles. *The Annals of Probability* **5**, 876-887.
Hall, P., Heyde, C.C. (1980). *Martingale Limit Theory and its Applications*, Academic Press.
Hansen, B. (1992). Convergence to stochastic integrals for dependent heterogeneous processes. *Econometric Theory* **8**, 489-500.
Jacod, J. (1975). Multivariate point processes: predictable projection, Radon-Nikodym derivatives, representation of martingales. *Probability Theory and Related Fields* **31**, 235-253.
Jacod, J. (1979). Calcul Stochastique et Problèmes de Martingales. *Lecture Notes in Mathematics* **714**.
Jacod, J. (1997). On continuous conditional Gaussian martingales and stable convergence in law. *Séminaire de Probabilités XXXI* 232-246.
Jacod, J. (2003). On processes with conditional independent increments and stable convergence in law. *Séminaire de probabilités XXXVI* 383-401.
Jacod, J. (2004). The Euler scheme for Lévy driven stochastic differential equations: limit theorems. *The Annals of Probability* **32**, 1830-1872.
Jacod, J. (2008). Asymptotic properties of realized power variations and related functionals of semimartingales. *Stochastic Processes and Their Applications* **118**, 517-559.
Jacod, J., Protter, P. (2003). Asymptotic error distributions for the Euler method for stochastic differential equations. *The Annals of Probability* **26**, 267-307.
Jacod, J., Shiryaev, A. (2003). *Limit Theorems for Stochastic Processes*, Springer.
Jakubowski, A. (1996). Convergence in various topologies for stochastic integrals driven by semimartingales. *The Annals of Probability* **24**, 1-21.
Jakubowski, A. (1997). A non-Skorokhod topology on the Skorokhod space. *Electronic Journal of Probability* **24**, 2141-2153.
Jakubowski, A., Mémin., J., Pagès, G. (1989). Convergence en loi des suites d'intégrales stochastiques sur l'espace D^1 de Skorokhod *Probability Theory and Related Fields* **81**, 111-137.
Kinght, K. (1991). Limit theory for M-estimates in an integrated infinite variance processes. *Econometric Theory* **7**, 200-212.
Kuelbs, J., Kurtz, T., Zinn, J. (2013). A CLT for empirical processes involving time-dependent data. *The Annals of Probability* **41**, 785-816.
Kurtz, T., Protter, P. (1991). Weak limit theorems for stochastic integrals and stochastic differential equations. *The Annals of Probability* **19**, 1035-1070.
Leadbetter, M., Rootzén, H. (1988). Extremal theory for stochastic processes. *The Annals of Probability* **28**, 885-908.
Lin, Z. (1985). On a conjecture of an invariance principle for sequences of associated random variables. *Acta Mathematica Sinica, English Series* **1**, 343-347.
Lin, Z. (1999). The invariance principle for some class of Markov chains. *Theory of Probability and its Applications* **44**, 136-140.
Lin, Z., Bai, Z. (2010). *Probability Inequalities*, Springer.
Lin, Z., Choi, Y. (1999). Some limit theorems for fractional Lévy Brownian fields. *Stochastic Processes and their Applications* **82**, 229-244.
Lin, Z., Lu, C. (1996). *Limit Theory for Mixing Dependent Random Variables*, Science Press/Kluwer Academic Publisher.

Lin, Z., Qiu, J. (2004). A central limit theorem for strong near-epoch dependent random variables. *Chinese Annals of Mathematics* **25**, 263-274.
Lin, Z., Wang, H. (2010). Empirical likelihood inference for diffusion processes with jumps. *Science in China* **53**, 1805-1816.
Lin, Z., Wang, H. (2010). On convergence to stochastic integrals. *arXiv:1006.4693.*
Lin, Z., Wang, H. (2011). Weak convergence to stochastic integrals driven by $\alpha-$ stable Lévy processes. *arXiv:1104.3402.*
Lin, Z., Wang, H. (2012). Strong approximation of locally square-integrable martingales. *Acta Mathematica Sinica, English Series* **28**, 1221-1232.
Lindbwerg, C., Rootzén, H. (2013). Error distributions for random grid approximations of multidimensional stochastic integrals. *The Annals of Applied Probability* **23**, 834-857.
Liptser, R., Shiryaev, A. (1983). Weak convergence of a sequence of semimartingales to a process of diffusion type. *Matematicheskii Sbornik* **121**, 176-200.
Liu, W., Lin, Z. (2008). Strong approximation for a class of stationary processes. *Stochastic Processes and their Applications* **119**, 249-280.
Lumley, T. (2013). An empirical-process central limit theorem for complex sampling under bounds on the design effect. Forthcoming in *Bernoulli.*
Marcus, M. (1981). Weak convergence of the empirical characteristic function. *The Annals of Probability* **9**, 194-201.
Mikosch, T., Resnick, S., Samorodnitsky, G. (2000). The maximum of the periodogram for a heavy-tailed sequence. *The Annals of Probability* **28**, 431-478.
Oksendal, B. (2003). *Stochastic Differential Equations*, Springer.
Politis, D., Romano, J., Wolf, M. (1999). Weak convergence of dependent empirical measures with application to subsampling in function spaces. *Journal of Statistical Planning and Inference* **79**, 179-190.
Phillips, P. (1988). Weak convergence of sample covariance matrices to stochastic integrals via martingale approximations. *Econometric Theory* **4**, 528-533.
Phillips, P. (1990). Time series regression with a unit root and infinite variance errors. *Econometric Theory* **6**, 44-62.
Phillips, P., Perron, P. (1988). Testing for a unit root in time series regression. *Biometrika* **75**, 335-346.
Phillips, P., Solo, V. (1992). Asymptotics for linear processes. *The Annals of Statistics* **20**, 971-1001.
Prohorov, Y.V. (1956). Convergence of random processes and limit theorems in probability theory. *Theory of Probability and Its Applications* **1**, 157-214.
Protter, P. (2005). *Stochastic Integration and Differential Equations*, Springer.
Raikov D. (1938). On a connection between the central limit-law of the theory of probability and the law of great numbers. *Izv. Akad. Nauk SSSR Ser. Mat* **2**, 323-338.
Rootzén, H. (1977). On the functional central limit theorem for martingales. *Probability Theory and Related Fields* **38**, 199-210.
Resnick, S. (1975). Weak convergence to extremal processes. *Advances in Applied Probability* **5**, 951-960.
Resnick, S. (1986). Point processes, regular variation and weak convergence. *Advances in Applied Probability* **18**, 66-138.
Resnick, S. (2007). *Heavy-tail phenomena: Probabilistic and Statistical Modeling*, Springer.
Skorokhod, A.V. (1956). Limit theorems for stochastic processes. *Theory of Probability and Its Applications* **1**, 261-290.
Song, Y., Lin, Z., Wang, H. (2013). Re-weighted functional estimation of second-order jump-diffusion model. *Journal of Statistical Planning and Inference* **143**, 730-744.

Sowell, T. (1990). The fractional unit root distribution. *Econometrica* **58**, 495-505.
Talagrand, M. (1987). Donsker classes and random geometry. *The Annals of Probability* **15**, 1327-1338.
Taqqu, M. (1975). Weak convergence to fractional Brownian motion and to the Rosenblatt process. *Probability Theory and Related Fields* **31**, 287-302.
Tyran-Kamińska, M. (2010). Convergence to Lévy stable processes under some weak dependence conditions. *Stochastic Processes and their Applications* **120**, 1629-1650.
van der vaart, A., Wellner, J. (1996). *Weak Convergence and Empirical Processes*, Springer.
Volný, D. (1993). Approximating martingales and the central limit theorem for strictly stationary processes. *Stochastic Processes and their Applications* **44**, 41-74.
Wang, Q. (2012). Martingale limit theorems revisited and non-linear cointegrating regression. Forthcoming in *Econometric Theory.*
Wang, Q., Lin, Y., Gulati, C. (2003). Asymptotics for general fractionally integrated processes with applications to unit root tests. *Econometric Theory.* **19**, 143-164.
Wang, W. (2003). Weak convergence to fractional Brownian motion in Brownian scenery. *Probability Theory and Related Fields* **126**, 203-220.
Wu, W. (2003). Empirical processes of long-memory sequences *Bernoulli* **95**, 809-831.
Wu, W. (2007). Strong invariance principles for dependent random variables. *The Annals of Probability* **35**, 2294-2320.
Yan, L. (2005). Asymptotic error for the Milstein scheme for SDEs driven by continuous semimartingales. *The Annals of Applied Probability* **15**, 2706-2738.

Index

Printed in the United States
By Bookmasters